Advance praise for

Adolescents' New Literacies with and through Mobile Phones

"Easily the most anticipated book of 2017 when it comes to understanding young people's steadfast commitment to learning on, through, and with their mobile phones. Julie Warner knows firsthand of which she writes."

—*Donna E. Alvermann, Distinguished Research Professor, The University of Georgia*

"Adolescents today are writing more than ever and in spaces that we never before knew existed. These new digital and portable writing spaces are as dialogic as they are distinct. So it seems today's youth hold the world quite literally in their hands. They palm the future of literacy, newer literacies, in ways that translate cultural practices of old into meaningful social practices anew. In this most important contribution to the new literacy studies, Julie Warner takes us into the life of youth through mobile phones, hashtags, and 'digital curation' practices that illuminate an exciting and critical digital world of words to which we should become more attuned."

—*David E. Kirkland, Director, New York University Metropolitan Center for Research on Equity and the Transformation of Schools, and Professor of English Education*

"Smartphone use is likely the most pervasive form of embodied digital capital for youth. They are constantly integrating and adding new elements to their repertoires from each other, the environment, and in the quietude of their beds at night. Their coded speech or digital discourse, though, is under constant scrutiny and threat for extinction when schools put digital discourse on silent or mute. Julie Warner understands this, and provides pedagogical strategies and roadmaps for teachers to mediate this threat by unsilencing, unmuting and unlocking this digital capital in the classroom. For every teacher who has perceived smartphones as disruption, this book is your answer for powerful and innovative ways to unleash coded speech as a motivator for learning. In fact, by the time you finish reading this book, selfies will take on an entirely different meaning in your life and in classrooms. Snap and type away!"

—*sj Miller, Deputy Director of Educational Equity Supports and Services, Metropolitan Center for Research and Equity and the Transformation of Schools, New York University*

"There is so much more to mobile phones than calling friends and family. Mobile phones are portals to diverse, engaging, and creative practices. Julie Warner invites us into this rich world of phoned-centered literacy practices with respect and wonder for the young people who enact them. Sometimes playful, sometimes affect-laden, sometimes intellectual, the literacy practices that young people apply in their use and enjoyment of cell phones prominently display their everyday multimodal and discursive competencies. Taking a landscape view of mobile phone knowledge work, *Adolescents' New Literacies with and through Mobile Phones* examines mobile phones across spaces; social mediation of phone content and images; chronotopic explorations; and digital curation as telling examples of younger generations' often invisible repertoires of practice and ways of knowing."

—*Jennifer Rowsell, Canada Research Chair, Brock University*

Adolescents' New Literacies with and through Mobile Phones

new literacies ¶

AND DIGITAL EPISTEMOLOGIES

Colin Lankshear and Michele Knobel
General Editors

Vol. 79

The New Literacies and Digital Epistemologies series
is part of the Peter Lang Education list.
Every volume is peer reviewed and meets
the highest quality standards for content and production.

PETER LANG
New York • Bern • Frankfurt • Berlin
Brussels • Vienna • Oxford • Warsaw

Julie Warner

Adolescents' New Literacies with and through Mobile Phones

PETER LANG
New York • Bern • Frankfurt • Berlin
Brussels • Vienna • Oxford • Warsaw

Library of Congress Cataloging-in-Publication Data

Names: Warner, Julie, author.
Title: Adolescents' new literacies with and through mobile phones / Julie Warner.
Description: New York: Peter Lang, 2017.
Series: New literacies and digital epistemologies; vol. 79 | ISSN 1523-9543
Includes bibliographical references and index.
Identifiers: LCCN 2016023382 | ISBN 978-1-4331-4407-3 (hardcover: alk. paper)
ISBN 978-1-4331-4408-0 (paperback: alk. paper) | ISBN 978-1-4331-4413-4 (ebook pdf)
ISBN 978-1-4331-4414-1 (epub) | ISBN 978-1-4331-4415-8 (mobi)
Subjects: LCSH: Composition (Language arts)—Study and teaching—United States.
Mobile communication systems in education—United States.
Literacy—Technological innovations—United States.
Education—Effect of technological innovations on—United States.
Social media. | Smartphones.
Classification: LCC LC151 .W29 2017 | DDC 808/.0420712—dc23
LC record available at https://lccn.loc.gov/2016023382
DOI 10.3726/b11221

Bibliographic information published by **Die Deutsche Nationalbibliothek**.
Die Deutsche Nationalbibliothek lists this publication in the "Deutsche Nationalbibliografie"; detailed bibliographic data are available on the Internet at http://dnb.d-nb.de/.

29 Broadway, 18th floor, New York, NY 10006
www.peterlang.com

For Aunt LaRita. I miss you.

TABLE OF CONTENTS

LIST OF FIGURES

FOREWORD

James Paul Gee

Mary Lou Fulton Presidential Professor of Literacy Studies
Regents' Professor
Arizona State University

Julie Warner's *Adolescents' New Literacies with and through Mobile Phones* is timely and important. It is timely because many young people's most pervasive literacy practices today seem trivial to many adults. It is important because these new practices can, in fact, lead to 21st Century skills and seriously compete, in depth and impact, with school-based literacy practices.

Warner discusses the composing practices of three young people who have struggled with school-based literacy, but who are avid digital composers. Let me give my own example, one that Elisabeth Hayes and I discussed in our book *Women as Gamers* (2011). This example speaks to why I believe Julie Warner's book is crucial today.

The Sims series of video games are the best-selling video games in history. In *The Sims*, players (over half of whom are girls and women) design, build, and maintain families, neighborhoods and communities. *The Sims* is a life simulator. However, some players do not just play *the Sims*, they also use it to write a form of graphic fan fiction. In their most basic form, *Sims* fan-fiction stories are created by taking screenshots during *Sims* game play. Players write captions under the screenshots, captions that take the form of a continuous narrative. A "writer" has to produce a multi-modal (words and images) "text" using digital tools (the game, graphic-design tools to modify the images, and sites for posting, reviewing, and critiquing stories).

A girl named Alex (she was fifteen when we studied her) is a popular *Sims* vampire-romance fan-fiction writer who has thousands of loyal and admiring readers. Alex writes her stories for fellow teenagers on an interest-driven Internet site devoted to *Sims* fan fiction. Below is what one teen has to say about Alex's *Lincoln Heights* series of stories. It is clear that for this teen, Alex's writing is helpful in a deep way (This quote is from the discussion thread on *Lincoln Heights* at *The Sims Writers' Hangout*):

> i LOVELOVELOVE Lincoln Heights. When i read it it always seems to cure my sadness and it has actually helped me deal with alot of depression and shit i've been forced to deal with lately. I can't wait until the next chapter is out… HURRY UP, ALEX!
>
> xo. :]

When Alex started writing stories on the site, she spelled poorly, her writing was weak, and she came close to plagiarizing the famous *Twilight* books. School held little for her in the way of literacy. Others on the site made it clear to her that posted stories had to be well-written, well-spelled, well-edited, and original. And they mentored her and helped her. She saw models of good work and received encouragement and critical feedback on her own work. She has now become adept at graphic design and writing. Indeed, some readers tell her they like her stories better than the *Twilight* stories.

Consider, for instance, the beginning of *Lincoln Heights*. As the story's description, Alex uses a quote:

> Description:
>
> "You cannot call it love, for your age the heyday in the blood is tame"—William Shakespeare

Then the actual story begins with (remember each of these sentences has a specially designed image accompanying it):

> Devin Collins rolled his dark eyes as his younger sister, Julia, whined about leaving all her little friends in New York.
>
> "I still don't get why 'we' have to move here when it's Ariel's parents who died!"

> Devin glared at his sister. She was ten: five years younger than himself, but she knew exactly how to piss him off.

Note here how "whined" and "little friends" are, though in the narrator's voice, actually infected by Devin's voice, as it is he that would think of Julia as whining and her friends as "little." This is a form of indirect quotation that is common in novels. The direct quotation "I still don't get why 'we' have to move here when its Ariel's parents who died!" is not explicitly attributed to the sister, but is clearly understood to be her speaking and, indeed, a representation of her whining. The emphasis on "we" gives the word a stress that sounds like whining.

The introduction of Ariel, a character we do not yet know, starts the story in "media res" (in the middle) and, again, places the narration inside the private knowledge and minds of Devin and his sister. This creates suspense and forces the reader to wonder who Ariel is. This, too, is a technique common in novels. And, again, "piss him off," though not a direct quote is an indirect quote of Devin's language or thoughts. And, of course, "rolling his dark eyes" is a phrasing meant to capture teenage girl readers at the outset (and it does—girls on the thread say Devin is "theirs" and comment on his attractiveness repeatedly). Alex, indeed, knows her readers very well.

Alex posts regular updates on her blog and on the SWH thread to keep her fans informed about progress on upcoming chapters, as well as other things about her life and writing. Consider one of these posts below:

> As u can see I gave my page a little makeover! I've had that oldd one for over a year! Needed a change! As 4 LH 1.3 I've got around thirty slides, working up to my usual 127! Patience is all it takes! I garentee it'll B out B4 Xmas though! ;)
>
> <3 A

Language like this gives rise to great consternation on the part of many adults. Some of them even wonder if such things betoken the death of written language as we know (knew) it. The spelling errors and unedited typos bother some people to no end.

However, we cannot conclude anything about Alex's post and its language unless we embed it in the whole set of practices and the larger activities in which it lives. Alex is engaged in a Discourse. She is enacting the identity of a *Sims* fan-fiction writer of vampire romance. This is an identity her teen readers readily recognize.

This identity involves engaging in actions that compose the larger overall activity of *Sims* fan fiction writing. This larger activity requires what I will call a "literacy tool-kit". Think of a carpenter. A carpenter carries around a tool-box full of tools, but also carries around a set of skills that allow the carpenter to put these tools to use in order to be a carpenter and do carpentry. So, too, with Alex. She has a tool-kit filled with tools and skills.

Here is just a part of Alex's tool-kit (tools and skills): She can use in-game design tools as well as various other design tools such as Adobe Photoshop. She has created and maintains her own personal website through which she energizes her fan base. She is an excellent writer and editor of her genre of graphic fiction. She knows well how to design images and match them to text (a master of multi-modality then).

Note that the post we saw earlier from Alex—the one that bothers some people for its errors and informality—plays a specific role in the ecology of what Alex is doing and who she is being when she is being a *Sims* fan fiction writer. It is a way to bond with her teen readers and to motivate them to be devoted readers and followers (her fans). Its language is meant to function in these terms and does so very well, indeed.

Note that Alex writes quite differently when she writes her stories. Here she edits them (with others) and they are free of "errors". This is because this form of writing plays a different role in the larger ecology of who Alex is being and what she is doing.

Such tool-kits can define for us an alternative notion of "transfer" of learning. As a person moves to new Discourses and new activities, it may be that the tool-kits they demand for mastery contain similar or related tools and skills as earlier Discourses and activities a person has mastered. We can ask: What is Alex better prepared to do—what Discourses and activities is she better prepared to learn—given the Discourse and associated tool-kit she has already mastered as a Sims fan fiction writer?

I am suggesting that one way to study old literacies and new ones, and their combinations, is to ask for any text, tool, or practice how it is situated within a given literacy tool-kit within the larger social world of Discourses. This goes for something as old as storytelling or as new as blogs, tweets, and mobile phones.

PREFACE

I took my first job as a teenager in the 1990s just so I could afford a beeper. Sometimes called a pager, having one of these little pieces of LCD-screen equipped plastic devices meant that you could communicate with your friends and family when you were out and about. The beeper came with a clip that allowed you to affix the thing to your clothing. Beeper communication was one-way; my beeper relied on input from a (usually landline) telephone to receive any information and I could not send messages through the beeper. When someone called your beeper number from a telephone, an automated prompt would ask them to input a callback phone number. While the beeper was designed to have the caller input a phone number, teenagers quickly implemented a system of codes that translated to simple messages. For example, inputting "143" meant, "I love you" (the numbers corresponded to the number of letters in each word in the phrase). Adding "911" after a phone number imparted a sense of urgency (call me back quickly!). "411" meant, "call me and give me the information on what is happening." Teens used numerals to suggest alphabetic text as well; 14 stood for "hi" because it resembled the word when the pager was turned upside down (the same went for 07734, which looked loosely like the word "hello"). My friends and I each had a numeric code that represented our selves. Mine was 791, chosen at random. These allowed us to sign our messages, since there was no way to indicate the identity of the sender built in to the pager interface. Beeper code dictionaries began to pop up online; I myself kept a decoder list to help me decipher some of the complex numeric messages I would receive. Such crib sheets were common among my friends, as well.

Perhaps it was this early, rudimentary form of text messaging that first piqued my interest in mobile messaging and the creative literacy practices of teens with technology. Although some of the provocative writings in recent years decrying the ways that smartphones are destroying culture and causing our social skills to atrophy frame the use of mobile technology as a new problem, I argue that adolescents have been engaging in creative and complex communication practices that involve short-form communications for decades. In a 1997 article in the Los Angeles Times, "Teens Create Language of Pager-Speak," the reporter writes,

> It's a secret language among friends.
>
> It may look like a jumble of numbers and asterisks, but it's actually a growing lexicon of the mundane, offbeat and obscene.
>
> While the language doesn't have a name, young people across the Southland and the nation rely on it to communicate those little messages that don't warrant a long conversation: "good night," "you're on my mind" or something decidedly less friendly.
>
> Welcome to the world of pager-speak.
>
> By dialing numbers that look vaguely like digital letters—right side up or upside down—the young linguists put together words and phrases.
>
> "They're on the cutting edge," said Michael Haddad, of Soft-Cell Communications in Beverly Hills. "They're the ones inventing the uses of the pager."

It wasn't until later when I graduated college ten years after my beeper days that text messaging became widespread; most cell phones did not have keyboards that supported fluid typing. In order to send a text message on the average cellphone, a number on the keypad would have to be tapped until the desired corresponding letter was recalled. This early text messaging was cumbersome and time consuming. Engaging in this early text messaging practice, I could not have predicted the kind of expanded multimodal toolkit with which teenagers would be able to communicate on-the-go across spaces with efficiency and directness (the beeper technology of my adolescence could hardly be thought of as a direct line of communication, for example!).

The objective of this book is to provide a deeper understanding of the phone-based composing practices of youth and their implications for literacy learning. In the United States, smartphone use among teens is nearly universal (Smith, 2013). Phone-based access shapes digital literacy practice primarily by way of the provision of the digital keyboard, which makes the typing of long alphabetic texts cumbersome, and the camera, which allows for communicating with images, an increasingly popular youth communicative practice. These realities inspired me to research the phone-based composing practices of three youth who struggled with formal schooled literacy but who were avid digital composers. The practices of three youth aged 14 and 15 will be presented in the pages that follow to explore phone-based composing as a social act involving relationships among multivocal texts, authors, and audiences. The widespread and rapid embrace of smartphones by youth from all income levels has had a substantial impact on the way that young people approach the act of composing, yet to date, little to no work has explored digital photography and text curation through popular apps like Twitter, Instagram, and Snapchat and their impact on literacy, including formal schooled literacy.

The data drawn from the case studies presented in this book illustrate the ways in which the participants' mobile phone based literacy practices, while highly individualized, shared several characteristics. Through the data collected I show the ways in which youth reconceptualize authorship through practices that include the voices of others through retweeting, remixing, and other kinds of text curation. Youth take into consideration feedback from audiences during the composing process in a way that shows audiences to be actively involved in the texts' authorship. Beyond harnessing alphabetic texts, youth also use the cameras in their phones to communicate. In their use of the cameras within their mobile phones, youth capture and sometimes stylize their physical worlds to communicate ideas without encoding ideas in alphabetic text. They use their faces, bodies, found objects, and physical worlds to communicate through selfies and other digital photography. The use of the built-in camera and apps that did not require logging in or other interruptions to fluidity of process allowed the youth to compose rich multimodal texts rapidly and spontaneously. Despite early deterministic claims about the Internet's damaging effects on language and thought, young people like the participants in my research demonstrate agency through ratifying, remaking, and remixing existing voices and practices in the composing spaces they enter with and through mobile phones.

The first part of this book provides the conceptual framework for the major issues explored in subsequent chapters. The opening chapter addresses the longstanding and ongoing debate around the ways in which technology influences society through some of the arguments that specifically deal with youth (e.g., deterioration of language, Google-induced stupidity). I lay out the concepts of tools, architecture, and affordances that underpin the analysis that follows. I outline the affordances of the mobile phone (e.g., mobility, camera) and the architecture of the apps that will be explored in later chapters (Twitter and Instagram). I explore the concept of encouraged practices (practices, I argue, engendered or afforded by the presence of particular architectures and affordances) and the ways that encouraged practices are subject to human interpretation. That chapter concludes with a description of some of the features of practices that the affordances and architecture of the mobile phone make possible. (Later chapters show the variance of use of these tools [i.e., human interpretation] in a way that challenge a technological determinist/essentialist approach to these concepts and solidify a social constructivist view of technologies and literacies). These include the youth practices that are explored in later chapters including social photosharing and text curation.

Chapter 2 further explores the conceptual landscape including the notion of the digital divide in an age in which smart phones are used by people as a means of affordable Internet access and where there is no difference in smartphone ownership between the highest and lowest socioeconomic groups. We know that youth composing with mobile phones are doing so as a mobile practice across all of the physical spaces they inhabit and in a variety of digital spaces. This reality has inspired literacy researchers to rethink previously dominant conceptions of space and place and what these mean for digital literacy research. In Chapter 3, I explore this notion of the spatial turn as it relates to mobile phone-based literacy practices. In Chapter 4, I draw from Bakhtin's dialogism (1982, 1986) to present a sociocultural view of composing with and through mobile phones. I present composing as situated, addressive, and social, and contrast it to the idea of composing as an individual accomplishment. I explain the ways in which youth use mobile phones for composing that show the production and reception of texts to be mutually constitutive.

Chapter 5 brings together the ideas of the first and second chapters to show the ways in which the affordances and architecture of the mobile phone and the engendered practices challenge previously-dominant views of composing as planned, designed, and focused on an end-product (cf. Leander & Boldt, 2013). In this chapter, I argue that instead mobile phone-based composing practices show composition as part of "ongoing discursive activity" (Vasudevan & Hill, 2008, p. 231).

In Chapter 6, I introduce the practice of the selfie (self-taken self-portrait), a practice engaged in by 91% of teens (Madden et al., 2013). An integral part of the selfie is the social, addressive component: the selfie is posted online, allowing others to comment on the photo and offer feedback, suggesting that selfies are a means for youth of garnering "feedback and affirmation" (Lenhart et al., 2007) as they develop their identities. Because mobile phone cameras offer a means of composing digitally without typing long messages (cumbersome to do on a small or digital keyboard), they represent a powerful means of communicating quickly and without friction.

Chapter 7 explores the ways in which digital curation is a practice that is easy to engage in with the mobile phone because so much semiotic richness can be accessed through text curation. I show how the retweeting function in Twitter allows composers to speak through others' voices with the press of a button (much easier than typing a long-form text with the digital keyboard). This chapter explains the ways in which this communication is layered: it allows the youth to show agreement with the idea as well as to signal alignment with or to affirm the original poster. I show the ways in which, in the practice of text curation, the practices of reading and composing are almost collapsed. Composers can read a text in their feed and then "grab" it and integrate it into their feeds. The boundaries between author and audience collide and blur in this practice. Composers are in dialogue with the texts they consume in this way as they subscribe to and then take in posts from their connections, reflect on what they believe and disbelieve from this world of ideas, and grab and display what they agree with or like in their own digital spaces. Youth also use curation (often through favoriting) as a means of holding onto digital content they want to revisit later, allowing for future reflection.

Chapter 8 argues that we need to go beyond a simple application of the critical digital literacies frameworks employed with print texts to take up more dynamic conceptions of critical literacies that consider the unique affordances of digital spaces for composing such as their persistence and replicability. In addition, I argue that when composing in a space wherein the dominant ethos is one of curation, being able to recognize the intersections of discourses and power is a literacy skill more important than ever before. Chapter 9 explores mobile phone-based composing practices through the lens of Bakhtin's chronotope to show how the participants represent time-space in their digital compositions.

The final chapter brings together some of the major ideas raised in the previous chapters and discusses their implications for literacy learning. Given this major shift toward social, collaborative, and dialogic composing amid a dominant paradigm of literacy that divorces composers and texts from their networked contexts, I explore the question of how those concerned with youth literacy can fruitfully build on the strengths and repertories that youth bring into spaces for schooled literacy. While schooled literacy has recognized that composing involves the integration of discourses (for example, persuasive writing as it is often framed by official curriculum, is highly addressive, and research writing is explicitly multivocal) composing is still framed as an individual accomplishment. I argue that capitalizing on the tendency of youth to make meaning intertextually and dialogically would benefit youth in the pursuit of school-sanctioned literacy learning. I suggest the ways in which activities involving curation and the creation of exhibitions and collections might be productively integrated into schooled literacy practice. I also suggest the ways that photography and social photosharing might represent a sophisticated way to communicate ideas in the schooled space without the "middleman" of linguistic text (in other words, representing ideas directly without encoding them in alphabetic text).

Most of the practices described in this book evoke issues that touched my life as a teenager who wrote with and through mobile technologies, as a classroom teacher and professional developer who specialized in technology integration, and now as I work to support education policy that reimagines the role of technology in literacy education. This text aims to support a better understanding of the rich and complex literacy practices in which youth engage so that responsive literacy education practice and policy can be developed.

References

Bakhtin, M. (1982). *The dialogic imagination*. Austin: University of Texas Press.

Bakhtin, M. (1986). *Speech genres and other late essays* (V. W. McGee, Trans.). Austin: University of Texas Press.

Gee, J., & Hayes, E. (2011). *Women as gamers: The Sims and 21st century learning*. New York: Palgrave Macmillan.

Leander, K., & Boldt, G. (2013). Rereading "A pedagogy of multiliteracies": bodies, texts, and emergence. *Journal of Literacy Research*, 45, 22–46. doi:10.1177/1086296X12468587

Lenhart, A., Madden, M., Smith, A., & Macgill, A. R. (2007). *Teens and social media*. Retrieved from http://www.pewinternet.org/Reports/2007/Teens-and-Social-Media.aspx

Madden, M., Lenhart, A., Cortesi, S., Gasser, U., Duggan, M., Smith, A., & Beaton, M. (2013). *Teens, social media, and privacy*. Retrieved from http://www.pewinternet.org/Reports/2013/Teens-Social-Media-And-Privacy.aspx

Mozingo, J. (1997, November 26). Teens create language of pager-speak. Los Angeles Times. Retrieved from http://articles.latimes.com/1997/nov/26/news/mn-57816.

Smith, A. (2013). *Technology adoption by lower income populations*. Retrieved from http://www.pewinternet.org/Presentations/2013/Oct/Technology-Adoption-by-Lower-Income-Populations.aspx

Vasudevan, L., & Hill, M. L. (2008). Moving beyond dichotomies of media engagement in education. In M. L. Hill & L. Vasudevan (Eds.), *Media, learning, and sites of possibility*. New York, NY: Peter Lang Publishing.

ACKNOWLEDGEMENTS

Thanks first and always to my mentor, Marjorie Siegel, for giving me the encouragement and direction needed for this work to take root and for being one of my most treasured friends. To John Broughton, who gave me the frameworks for thinking about composing with technology, and to Chuck Kinzer, who pushed me to think about why this work matters. To Sheridan Blau, who offered support all along the way, even long before I arrived at Teachers College.

Of course, this inquiry would not exist if not for the generosity of Ms. Harrison, who opened her classroom to me and sparked my thinking about teaching with technology and indeed teaching in general. Much love to you for letting a doc student add one more thing to your already overflowing plate.

Likewise, to the three young people who agreed with no reservations to take part in this research; my deepest gratitude goes out to you.

Michele and Colin, thank you for believing that this work was important enough to find a place in this series, which I read before even entering my doctoral studies and continue to read with the same sense of curiosity and delight. To write this book is a dream come true.

To Kurt Morris for always generously offering substantive reads and edits that allow me to get my ideas out in the world.

To David Cook for sparking my thinking about new media and for being willing to work through ideas with me over text message!

To Heather Pleasants, Ryan Rish, Damiana Gibbons Pyles, Lindy Johnson, Darren Crovitz, and Sean Connors, thank you for the engaged scholarship that shaped in so many ways my thinking and the work in these pages.

To Lesley Roessing and the rest of my Coastal Savannah Writing Project family, thank you for always keeping me grounded in what this work is really all about—what teachers and young people do every day.

And finally, to my family. I love you.

· 1 ·

OVERVIEW OF THE STUDY

Abstract

This chapter discusses the methodology used in the study of three adolescents composing with mobile phones, connective ethnography. The chapter begins by providing background on the connective ethnographic orientation to researching mobile literacies, and presents the participants, research context, and methods of data collection that account for mobility in and across online and physical spaces. Finally, the chapter explains the methods of data analysis employed in this study of mobile literacy practices.

Keywords: Connective ethnography, mobility, mobile literacy practices, methodology, research context, adolescents, mobile phones, composing

The data presented in this book comes from a fourteen-month connective ethnographic (Hine, 2000; Leander & McKim, 2003) study of the phone-based digital literacy practices of three adolescents across physical and digital spaces. The data gathered through the process of this research, which involved observing the participants' interactions in online spaces, speaking at length with them about how they thought about their composing practices, analysis of the texts they composed, and engaging in these spaces as a participant myself, allowed me to build a nuanced understanding of the complex and highly inventive communicative practices of youth who compose with mobile phones. This research constitutes the foundation for this book, and the book explores the themes extending from that research context.

The networked and mobile digital media at the heart of this work have problematized existing notions of place and space; they have become unbounded and destabilized. McCullough (2004) maintains, "Digital technology now pours out beyond the screen into our messy places …" (p. 6). In order to deal with the issues of place and space ("the field") in this ethnographic study centered on mobile digital media engagement, I employed connective ethnography (cf. Hine, 2000; Leander & McKim, 2003), which considers the offline and online as part of one continuum. Connective ethnography is taken up by researchers who seek to describe and interpret participants' practices both online and offline through and across the spaces they inhabit (e.g., boyd, 2008; Hine, 2000; Markham, 1998). Researchers have attested to the "seamlessness" of young people's online and offline realities (e.g., boyd, 2008; Thomas, 2007), and research shows that teenagers' self-presentation and written expression on the web is in harmony with their offline lives (Blinka & Smahel, 2009; boyd, 2007, 2008; Huffaker & Calvert, 2005; Stern, 2007). In carrying out this study, I called upon the traditional tools of the ethnographer: participant observation, field notes, artifact collection, and interviews across digital and physical spaces. I used participant-kept logs of online travels and hyperlinks within participant web spaces to follow participants' interactions online, as well, in order to document as much of these three young people's everyday social practices as possible in order to better understand their uses and value for the participants within the scope of their everyday lives. These are discussed in more detail later in this chapter.

Ethnography (literally, "writing culture") extends from anthropological approaches to research and was, at first, a way of studying a culture that was usually tied to a particular geographic location or focused on one ethnic group or culture with a shared geography. More recently, ethnographic approaches have been used to study cultures that are not defined by geography or ethnicity, such as cultures of particular organizations. Technology has made geography even less relevant as a means of making groups or cultures cohere. Accordingly, connective ethnography has moved away from ethnography's focus on cultures and subcultures and instead facilitates exploration of social practices more broadly. In mobilizing an orientation to research rooted in connective ethnography, I was able to study how technology functioned as a part of the participants' everyday social lives. For the connective ethnographer, the online is both a space for interaction where one can study human interaction and its social meanings, but it is also an artifact, in some cases related to "in real life" (IRL) or offline action with concomitant social uses and interpretations.

Participant Selection and Research Context

I never set out to conduct a generalizable or representative study, but was instead interested in gaining a rich understanding of what young people who struggle with schooled literacy do in their lives online and offline with mobile phones. For this reason, purposive sampling (Merriam & Tisdell, 2015) was used to identify the research site, White Birch High School (all names of institutions, places, and participants are pseudonyms), a Title I (a program providing funding to local educational agencies or schools with high numbers or percentages of students from low-income families) "failing" technology magnet high school within the Southeastern United States, to represent a focal population that engaged in digital literacy practices with a history of failure in schooled literacy. In the United States education system, a magnet school is a public school with a specialized emphasis, and in this case White Birch High curricula were focused on technology. The school opened in 2010 with the aim of reducing the dropout rate in one of the most economically depressed areas in the county in which it is situated through the integration of technology including science labs and iPads. White Birch High focuses on technology integration to break the "cycle of failure," as a school board chairman put it.

In undertaking data collection, I carried out initial observations for six months in a ninth grade English class of a cooperating teacher, Ms. Harrison, and then followed three focal participants in the online spaces they frequented for the remaining eight months of the study. For the duration of the study, I moved back and forth among online and offline spaces, collecting data online and offline and being careful not to privilege the online or the offline as sites for data collection.

Within that class, I sought out participants who I could follow for the duration of the study and who would offer insight into the meanings they made of their own practices. Purposive sampling (Merriam & Tisdell, 2015) was thus also used to find focal participants who were frequent and skillful users of technology from the ninth grade class. I used a brief questionnaire administered to all students in the ninth grade class that asked about students' composing practices, frequency of use, and tools for Internet access. Selected focal participants included three students; they happened to reflect the gender and racial distribution of the class. All three of the focal participants accessed the Internet primarily via mobile phones. This was found to be typical across Ms. Harrison's ninth grade English class. All three participants reported that they were heavy users of the social media platforms Twitter and Instagram. The participants were on average 14 years old.

Focal Participants

The three focal participants in this research were Eli, John, and Aaliyah. They were all students in Ms. Harrison's ninth grade English class and all lived in the community around White Birch High School. While they accessed the same platforms online and used similar tools for composing, their practices were surprisingly individualized and offered rich grounds for a broad exploration of the mobile phone-based composing practices of youth and particularly of youth who struggled with schooled literacy practice.

Eli was 14 years old when this study began. Of Puerto Rican descent, he was one of the few non-African American students at White Birch High. Ms. Harrison described him as one of her brightest and best students. In the classroom space, he was always engaged, adding to class discussions and responding to questions posed by his teacher in a way that suggested he was invested in the freshman English course. Still, he failed the government-mandated End of Course Test. Eli was 10 years old when he started going online. He had been using the Internet, then, for four years when the study began. I asked Eli what he would do online at 10 years old, and he said, "Nothin. Act stupid, post pictures of me with my lunchbox. I found my MySpace [profile page] a couple years ago and I seen me when I was like 10 years old." He laughed at the thought of his early, "rookie" social-networking practice. Eli continued that social networking practice as a teenager, and posted tweets and photos to Twitter and migrated to Instagram where he posted (usually captioned) photographs. The phone was his preferred method of accessing these platforms.

The first day I observed Ms. Harrison's class, John immediately captured my attention as he very vocally participated in class, often "cutting up" in the words of his teacher, but more often contributing insightful feedback that moved classroom conversation in ways that led Ms. Harrison to describe him as one of her best students. He could make the whole room laugh or provoke deep reflection in equal measure. Like Eli, though, he performed poorly on the government-mandated End of Course Test. What I came to discover was that John was a prolific digital writer who posted frequently to Twitter with a large base of "followers," those who subscribed to his Twitter updates. John seemed to have a sense of how to produce content that was meaningful and relevant to his followers, as his posts were often retweeted and favorited by others. ("Favoriting" is conventionally a form of expressing appreciation for content on Twitter. It is also a means by which one can bookmark a tweet for later recall). In fact, one of the other participants in the study told me that John was "really good at Internet stuff."

In my first interview with John, I asked him about his history of digital composing.

> School taught me how to use the computer. Like, we had to go to the computer lab and stuff. You know Computer Labs Class at [name of middle school]? They had Computer Labs Class. We would, like, have to go on a learning site, they would have the site on the board, we would have to type it in and stuff. I would go home and I'd be like, Mom, can I play a game? And she'd be like, let me show you how to do it, and I'd be like, no I know how to do it! And type it in myself.

John carried these early technical skills learned originally in school into his out-of-school practice, and moved from a consumer of content, navigating to websites and playing games, to a producer of content. Like Eli, John used his phone as his primary means of accessing the Internet, and engaged in microblogging, social photo sharing, and retweeting. He told me, "I get on a lot in school, in church, and in the car, or at home."

Fourteen-year-old Aaliyah was just a freshman but had already attended three different high schools by the time I met her at White Birch High School. Rather than complain about having to continually uproot, she matter-of-factly described to me the ways that she harnessed mobile technologies, mainly Twitter, to keep in touch with her friends, dispersed among several schools. She created a space for herself and her friends, apart from her parents and family members (who were on Facebook, which she avoided for that reason), where she could be "stupid" and "post stupid stuff." Aaliyah understood her capacity to produce a new space with digital tools (Lefebvre, 1991). To access Twitter, she always used her phone. For Twitter, "I don't get on the computer. I don't like it on the computer," she told me. She explained further that she enjoyed the ease of access made possible by the Twitter app on her phone that meant she never had to stop and enter her credentials to log in.

I asked Aaliyah about what she might post on her Twitter page. She said, "Mostly when funny things happen, or when I have to talk about something." She then added she might also post when she is "doing something stupid," which she explained meant something "corny, like a joke or something." As she explained it, the Twitter space was a space for her and her friends to joke around. "If we were to talk about deep stuff," she explained, "they would direct message me or we would talk about it on the phone." Indeed, Aaliyah's expressiveness varied in different spaces. At school, she was quiet and reserved. Online, she was boisterous and silly. Her online handle or screen name on Twitter included derivatives of the words "psychotic" and "crazy." She saw Twitter as a safe space where she was among friends and where she displayed and posted content that reflected her personality, beliefs, sense of humor, and worldview.

In the pages that follow, I draw from the mobile phone-based literacy practices of these three young people to describe the ways adolescents think about and approach the act of composing using the phone with its particular affordances and constraints, to illustrate the ways in which the traditional idea of writing process is challenged, curation is a new form of authorship, and the ways youth rely on the voices of others as a central means of communicating with texts. The connectivity that adolescents harness through the mobile devices that accompany them across the different spaces they inhabit in their everyday lives has implications for literacy learning that can no longer be relegated to the periphery and must, as the Internet has in our classrooms, come front and center. This will prove to be particularly important for students who struggle with schooled literacy, as the expanded modal and semiotic toolkit that digital composing tools offer can support a broader array of learners in their expression than can linguistic text alone.

The next section offers more detail about the physical research site, including some of the characteristics of the local town, Brookeville, in which the participants lived and attended school at White Birch High School. It also describes the cooperating teacher who allowed me to conduct research in her classroom.

Research Context

White Birch High School is located in Brookeville, a small town in a rural area in the southeastern United States rife with majestic oak trees and pristine painterly landscapes. The river Rill meanders through the small town. The serenity of this place belies the destruction that took place during the Civil War era, although the ruins of historic buildings, many of the plantations, and the fishing and farming traditions of that period remain. The aftermath of those times is manifest in the stark contrast between the wealthy and those relegated to living in poverty in Brookeville. Resorts and gated communities line the river, while a quarter of the people in Brookeville live below the poverty line, mostly in the wooded areas off the main highway that guides traffic to nearby larger cities. Brookeville residents live in relative isolation; the town is reachable only after traversing the river and then that long stretch of highway, with nothing but acres of farmland visible on either side of the road. Tucked away a mile or so off the highway are those zoned for White Birch High School, living in mobile homes, several to a lot. Once turning off the highway toward the school, cell service drops, phones reading "NO SIGNAL." The economy here has capitulated to the surrounding towns. What exists in Brookeville is predominantly agricultural, and residents have to travel outside of Brookeville to shop for things like clothing, furniture, or even groceries, as the only buildings in the downtown area are the tiny post office, a few gas stations, overpriced convenience stores, and cafes that serve food from attached farmland.

In collecting data for this study, I worked with Ms. Harrison, a 38-year-old African American woman who taught English at White Birch High School. Ms. Harrison was herself local to the area, and had attended the other high school in the district, a school that White Birch High was built to supplement. At the time I began this study, Ms. Harrison had been teaching for 13 years, and had been at White Birch High since it opened. In the opening years, the school experienced an extremely high teacher turnover rate (including almost 100% in the English department), positioning her as one of the most experienced and rooted teachers in the school; she was chair of the English department. I inquired about Ms. Harrison's ideas about why there was such a high turnover rate in the department, and she explained that she thought it had to do with a culture clash of sorts. Teachers were coming to teach at White Birch High "tricked" by the early college distinction into thinking that the high school would serve a "different population," in her words. She suggested that because she is a part of the local community and has a rapport with parents, her approach to classroom management is more effective: "I can just tell [the students], I'm going to take you and put you in the car and take you over to your mom's house." The teachers who left were White, while the population of the school and surrounding community is predominantly African American. At White Birch High, 76% of students are African American, 18% are White, and 4% are Hispanic.

Knowing that I was interested in working with her particularly because of her experience and education in teaching with technology, in our first meeting Ms. Harrison described the trouble that technology caused for the business of teaching. All the students "want to use their resources," she said. In an age of ubiquitous computing and "just-in-time" information, students no longer see the value in committing information to memory. Because of this, she explained the ways in which she has had to change the way she teaches. For example, sometimes she will let students use iPads on their tests and refer back to their notes. To ensure that students are not merely engaging in low-level searches and regurgitation, she created application questions for tests.

To Ms. Harrison, the idea of the digital native is a total myth: "All they know how to do is Internet-type stuff." When she tried to bring in a new technology-enhanced lesson,

> it ends up being like I'm teaching a computer class because they don't know how to do anything. Facebook, Twitter, and Google. That's what they know how to do on the computer.

Over the course of my observations, I noted that Ms. Harrison constantly had to impart technical knowledge through direct instruction as she was delivering her carefully crafted lessons. She took a great deal of classroom instructional time explaining how to open applications, save files, and perform other technical functions with the hardware and software that had nothing to do with content area material.

Ms. Harrison has several degrees, including an advanced degree in educational technology. However, she is quite humble, and downplayed her skills and knowledge when I asked her directly about her education. She claimed that none of the preparation she got from her master's program in educational technology is relevant anymore: "It's nothing I'd put on a resume." She said if she were to list that degree on a resume, an employer is going to expect her to be skilled in educational technology. However, the digital landscape has changed so much in the few years since she got the degree that she would be expected to do things she does not know how to do because what she learned in the program is now irrelevant and much of the technology obsolete.

Ms. Harrison's perspective around teaching was colored by her relationship with her daughter. Following her mother's example, Ms. Harrison's oldest daughter had just begun her first year of college studying to be an English teacher at the beginning of this study. The generational commitment to English-language-arts education in the family became, for her, a study in the changing landscape of the field. In talking to her daughter about a literacy education course she was taking, Ms. Harrison was surprised to find she had not asked her to edit any of her papers so far this year. Her daughter replied, "We haven't written any papers, Mom." Instead, the class had been working with GarageBand—a software application that turns a computer or device into a recording studio—and engaging in other "digital literacies." Ms. Harrison replied to her daughter, "I guess writing is a dying art!" As an English teacher in a technology-infused high school with a daughter preparing to educate the next generation of learners in the next four years, her vantage point is conversant with the idea of literacies as multiple and multimodal. However, she expressed frustration at the ways that she was constrained by the realities of a large-scale implementation of technology as new as iPads were.

I myself had stopped teaching in the United States public school systems just as digital media and technology tools had begun to be taken up by teachers. In my last year of teaching, the 2007–2008 academic year, I had tried to harness technology to enhance my teaching and was met with opposition from administration. Media discourses around technology at that time were fear-laden, evidenced by *The New York Times* articles from 2007 such as "New Scrutiny for Facebook over Predators" and "States Ponder Laws to Keep Web Predators from Children." Coming back into the classroom to work with Ms. Harrison, I was intrigued by the shift that had taken place over just a few years; during the time that I was in New York taking courses for my doctoral program, it seemed there had been a complete reversal in attitudes toward technology. Technology was now viewed as a panacea instead of a threat. Ms. Harrison's technology-infused teaching offered me a chance to see in action what I had hoped to enact in my own classroom but was blocked from performing.

Researcher's Role

Traditionally, the "trustworthiness" of a study has referred to the extent to which the findings are "true" and "objective." It is often used interchangeably with validity, rigor, or credibility. Postmodern times complicate conventional ways of thinking about trustworthiness. For example, ethnographic researchers assume the social world they study is a world of multiple realities rather than a single reality that exists apart from their knowledge of it. As a result, they acknowledge that they themselves are central instruments with particular positionalities that correspondingly affect data collection and analysis. The trustworthiness of a study, accordingly, is paradigm-dependent as well as researcher-bound. In other words, particular research traditions lay the groundwork for orientations to the concept of trustworthiness. As an ethnographer within an emerging tradition of connective ethnography, I see as transparent a description as possible of my positionality as a researcher as central to the trustworthiness or quality of the research. My role in this study was as a participant researcher and a co-learner. Even though as the daughter of a software developer and as an early adopter of mobile digital composing I had long-term familiarity with digital composing, I was unfamiliar at the start of the study with some of the tools the participants were using such as Instagram and had to install them on my phone to gain familiarity with the software's affordances, mechanisms, and social uses as I undertook the research.

I also see it as important that I explain that as a former high school English teacher in a Title I school, I had questions rooted in my experience that led to this study. That fact was not inherently problematic, but assumptions that arose from my background as a high school English teacher and my position as an instructor of composition at a university, my job at the time, were important to confront, including those regarding the texts and practices of English teaching in general, but also English teaching in higher education. Throughout the study I had to confront what texts, practices, approaches, and worldviews I privileged and interrogate my own assumptions about literate practice. Furthermore, as I now write this book, which emerges out of that research project, I explore the themes that surfaced through the lens of my current position as a researcher within the education policy planning and evaluation arm of the United States federal government. My current position inescapably colors my worldview and my view of the data.

I was limited in my perspective as an outsider in data collection. To put it another way, my positionality as a reader of youth practices positioned me outside the action by definition. My outsider status limited my ability to interpret how participants used the creation of texts and the texts themselves and their movement through space(s) to mediate their respective experiences. Also of note is my positionality as a white, middle-class woman, because I studied participants who were different from me and I had to be conscious of the assumptions I made about them. I took some steps to mitigate these realities.

Throughout this study, I attempted to make clear my preoccupations and values and to connect my prior experiences and background to my analytical perspective through analytical memoing (described in the section that follows on data analysis). In addition, to ensure that I was true to the experiences and perspectives of participants, I used two or more sources of evidence to draw conclusions related to my inquiry. For example, if a conclusion was drawn from considering an observation and interview responses in conjunction with one another, that conclusion is trustworthier than a conclusion drawn from one of those data sources alone. Whenever possible, three data sources were used to draw conclusions.

In addition, I attempted to document the participants' practices in ways that were faithful to the practices in situation. Because I transferred digital artifacts to print, I attempted to maintain awareness of what was lost in the transfer by making notes about where hyperlinks led. I used a two-column format for field notes that helped me to separate my own impressions from the data as much as possible. I used field notes to capture my thoughts and perspectives alongside more straightforward accounts of what happened in various spaces. As someone who grew up around technology, I was aware that my familiarity bred assumptions and that I needed to consistently check with participants as to whether or not my assumptions were in line with their perspectives. At several points throughout this study, I was surprised by the complexity of the participants' literacy practices after probing my assumptions with participants that would have eluded me otherwise.

Participant Observation Online

The process of following the research participants so that I could observe their composing practices was complex. First, I had to send requests to follow the participants based on the Internet pseudonyms (handles) they gave me. Often a mix of numerical and alphabetic characters that spelled out slang terms or snippets from song lyrics, these handles were difficult for me to replicate in the search functions online. Normally I type a name into the search function, and am able to find the person I seek; with the participants, however, this was nearly impossible because they did not link their real names to their online profiles and used complex handles with inventive formulations. In addition, they changed their names like someone might change their clothes. I repeatedly had to search through my feeds, scrolling manually to locate the participants' pages after they would change their handles to something more of the moment.

There were some spaces to which I did not have access in this study. Logistically, I could not follow the participants at all times, and my physical contact with them was limited to the school or if I bumped into them in the local town. Originally, I had wanted to carry out the final interviews with students in their primary physical composing spaces. Early in the study, I realized participants did not have a primary physical composing space. Instead, they interacted online in transit, in cars, at school, at home, and all the other spaces they inhabited.

Likewise, there were some online spaces to which I did not have access. Although participants did not use privacy settings on their sites and therefore most of their online interactions were visible to me, some interactions, like peer-to-peer digital interactions such as text messaging and Kik, a smartphone messaging service participants used to engage in group chats, it would have been intrusive for me to ask to observe there. I did not ask for access to Kik because I sensed it was a very personal space, akin to the space of text messaging between individuals. Sometimes, however, participants would post screenshots of interactions on Kik to their Instagram accounts, and got a sense of the nature of the interactions on Kik.

My observations then were limited to the more public sites of interaction. Although technology-supported peer-to-peer communication is important to consider, the online posting of digital compositions to intermediary sites rather than particular people was my original focus. In other words, there is a point of dissemination: a tweet is posted online, on a website, rather than sent directly from one phone to another, for example. This intermediary creates a space to which people might navigate to see and eavesdrop on the conversation and maybe add to it. It is public in nature and was the focal dynamic for this study.

Although many digital literacies researchers are taking up electronic tools to aid in data collection, I used a paper notebook with a two-column format for collecting field notes during both online and offline observations. I found that this was the least obtrusive way to take field notes in the school, and it kept me organized to keep all of my field notes in one place. In one column, I aimed to capture the elements of the environment and the social practices; I described in this column only what I observed. In the second column, I captured my interpretive notes and my emerging analysis. These perceptions were often the basis for analytical memos (described below).

Although a wide range of data was gathered, my analytical focus was on the data relating specifically to composition (defined as text creation or text dissemination) with mobile phones. The digital texts created by the participants varied across platforms and participants. For example, all three of the participants composed microblogs called tweets on Twitter, but the tweets varied across participants. For example, one participant seldom included images in his tweets, while another participant relied heavily on visual images in her microblogging practice. On a different platform, the social photo sharing site Instagram, only two of the participants composed using original photography, and both used original photos, curated images, and linguistic text captions to compose there.

Data Analysis

Researching mobile phone-based literacy practices means being prepared for a digital deluge. Because of the large amount of data collected (most of the participants created phone-based digital compositions daily), I captured data digitally and printed and filed it chronologically in a notebook on a near daily basis. I coded the data printouts by hand as I collected and formatted the data. Although it admittedly felt odd to turn digital data into print texts in a study of digital literacy practice, I found it most useful to capture data digitally and to analyze data by hand and on paper. I was able, using this format, to organize and reorganize data as I saw fit at different points in the study, and I was able to use sticky notes for coding. It also allowed me to keep data memos with corresponding data. Using Saldaña's (2013) model for coding data, I carried out first cycle coding, second cycle coding, and memo writing. In the first cycle, I coded the data using descriptive coding with my research questions as a guide and labeled data "practices" where I saw text creation or dissemination (i.e., composing). After I read, re-read, considered, and re-considered the data, I refined the first cycle codes to reflect recurring themes and patterns in the practices of the participants (Saldaña, 2013). For example, the code "practices" was refined in various cases as "curation," "photography," and sometimes even further as "favoriting as curation" or "deleting as part of composing."

Throughout data collection and coding, I followed Saldaña's (2013) memoing practice. Saldaña (2013) recommends writing memos in the field during data collection and while coding:

> The purpose of analytic memo writing is to document and reflect on: your coding process and code choices; how the process of inquiry is taking shape; and the emergent patterns, categories, and subcategories, themes, and concepts in your data—all possibly leading toward theory. (p. 32)

I used analytic memoing to explore my emergent ideas related to the participants' digital practice and to interrogate some of my assumptions about "legitimate" literacy practice. It complemented the coding process by bringing my ideas around the pithy codes into sharp relief. I also recognized a need early on to write about the status of the study's methodology to clarify potential shortcomings. For example, early on in the study, I noticed that participants were posting and then deleting texts regularly. This led me to integrate an iPhone as a tool for data collection as I identified the need for heightened vigilance as a participant observer. (My iPhone had more persistent access to the online space via a 4G connection than did my iPad, which required a wireless Internet connection to function fully and therefore could only be used where wireless Internet was available). I revisited these methodological memos during data analysis as well since they related to the youth's repertoires of digital practice.

Researching Mobility

In order to explore the geographies and other spatial qualities of the composing practices of youth accessing the Internet with mobile phones, I moved across digital and physical spaces in collecting data, beginning in the classroom of a cooperating teacher and later moving into the online spaces created or frequented by three focal participants. Early on in conducting the research, I found that I had to be as mobile as the participants and just as "always on" (Baron, 2008) as they were or I would miss important data sources; for that reason, I used mobile tools to collect artifacts like postings on social media, conversations, and multimodal digital compositions so I could collect data across places and spaces. Using the screen capture function on an iPad and an iPhone and remaining vigilant in observing their interactions in online spaces, I was able to collect a rich data set that was impossible otherwise, since the participants often deleted postings after a short period of time. Finally, I relied on the participants' accounts of their online travels to follow them in the online space as well as the geolocation feature on one of the social media platforms two of the participants utilized (Instagram). This feature provided information as to where the participant was physically located when photos were posted.

Interviews in this study took place both online and offline; as I gathered artifacts and observed the participants' engagement with digital media, I asked follow-up questions using the messaging functions on the platforms and email. The online interviews allowed me to link to texts and to communicate with participants quickly. I complemented the online interviews with face-to-face interviews so as to avoid privileging the online or offline during data collection, a choice in line with connective ethnographic methods. Face-to-face interviews allowed for more substantial responses from participants. For example, based on their responses and the direction of the conversation, I could follow up with prompts. In interviews, I asked questions related to the networked, multimodal, interactive, or global nature of digital media. Further, I raised questions about commonalties among practices I noticed as the study progressed, such as how participants manipulated privacy settings, and what affordances were available for composition.

To examine mobility, data were analyzed first through an extensive coding process based on Saldaña's (2013) two-tier coding heuristic wherein I coded in a first cycle for *spaces*, *connections*, *resources*. In a second cycle, I used theoretical coding (Saldaña, 2013), assigning a word or short phrase related to issues of space and emerging themes: *movement/mobility*, *control/creation of space*. I then used Inspiration software to map the space (i.e., connections among the social connections, places, resources) to attend to the research question and attempt to make visible the geographies of the focal participants' digital engagement using mobile phones that allowed them to move across places and spaces. This mapping was based on Saldaña's (2013) memoing technique wherein he suggests qualitative researchers "reflect on and write about the possible networks (links, connections, overlaps, flows) among the codes, patterns, categories, themes and concepts)" (p. 36). The objective is "rigorous reflection on the data" (Saldaña, 2013, p. 41) as well as a means toward theory-building. This visual memoing technique also afforded a means of visualizing space as a "field of relations" (Olwig & Hastrup, 1997, p. 8).

The themes generated through this analysis are explored in the subsequent chapters of this book. While I explore broadly practices and social uses and meanings of mobile phone-based literacy practices, most of the ideas in the following pages stemmed from the analysis of the data collected as I conducted observations in Brookeville and the digital spaces in which Eli, John, and Aaliyah interacted. They are also shaped by the many interviews conducted both online and offline with the three participants, who engaged in inventive and multifaceted communicative practice with and through their mobile phones.

References

Baron, N. (2008). *Always on: Language in an online and mobile world*. New York, NY: Oxford University Press.

Blinka, L., & Smahel, D. (2009). Fourteen is fourteen and a girl is a girl: Validating the identity of adolescent bloggers. *CyberPsychology & Behavior, 12*, 735–739. doi: 10.1089/cpb.2009.0044

boyd, d. (2007). Why youth (heart) social network sites: The role of networked publics in teenage social life. In D. Buckingham (Ed.), *Youth, identity, and digital media* (pp. 119–142). Cambridge, MA: MIT Press.

boyd, d. (2008). *Taken out of context: American teen sociality in networked publics* (Unpublished doctoral dissertation). University of California, School of Information, Berkeley, CA.

Denzin & Y. S. Lincoln (Eds.), *The Sage handbook of qualitative research* (3rd ed., pp. 507–535). Thousand Oaks, CA: Sage Publications.

Hine, C. (2000). *Virtual ethnography*. London: Sage.

Huffaker, D. A., & Calvert, S. L. (2005). Gender, identity, and language use in teenage blogs. *Journal of Computer-Mediated Communication, 10*. doi:10.1111/j.1083–6101.2005.tb00238.x

Leander, K., & McKim, K.K. (2003). Tracing the everyday "sitings" of adolescents on the Internet: A strategic adaptation of ethnography across online and offline spaces. *Education, Communication & Information, 3*, 211–240. doi:10.1080/14636310303140

Lefebvre, H. (1991). *The social production of space*. Cambridge, MA: Blackwell.

Markham, A. (1998). *Life online: Researching real experience in virtual space*. Walnut Creek, CA: Rowman Altamira Press.

McCullough, M. (2004). *Digital ground: Architecture, pervasive computing, and environmental knowing*. Cambridge, MA: MIT Press.

Merriam, S. B., & Tisdell, E. J. (2015). *Qualitative research: A guide to design and implementation*. San Francisco, CA: Jossey-Bass.
Olwig, K. F., & Harstrup, K. (Eds.). (1997). *Siting culture*. New York, NY: Routledge.
Saldaña, J. (2013). *The coding manual for qualitative researchers*. Los Angeles, CA: Sage.
Stern, S. (2007). Producing sites, exploring identities: Youth online authorship. In D. Buckingham (Ed.), *Youth, identity, and digital media* (pp. 95–117). Cambridge, MA: MIT Press.
Thomas, A. (2007). *Youth online: Identity and literacy in the digital age*. New York, NY: Peter Lang.

· 2 ·

CONCEPTUAL LANDSCAPE

Abstract

This chapter provides the framework for the major issues explored in the chapters that follow, beginning with addressing the longstanding and ongoing debate around the ways in which technology influences society through some of the arguments that specifically deal with youth culture (e.g., deterioration of language, Google-induced stupidity). It lays out the concepts of digital tools, architecture, and affordances with regard to digital literacy practices, and outlines the affordances of the mobile phone (e.g., mobility, camera) and the architecture of the apps that will be explored in the later chapters such as Twitter and Instagram. This chapter presents a broad review of relevant literature. It closes with an exploration of the current education policy landscape as it relates to to technology and in particular the intersection of policy, technology and rural schools.

Keywords: Digital architecture, affordances, digital tools, Twitter, Instagram, curation, social photosharing, youth culture, education policy, rural education

Technology integration in schooled spaces for literacy has changed drastically since the early days of the Internet. When I was a high school student in the 1990s, the dawn of the Internet, one could only access the Internet at school on the lone connected computer, sequestered in its own room in the library. Students had to sign a release form before they were allowed to use the slow dial-up connection and the waiver dictated that the Internet was to be accessed for research purposes *only*. Students could use online resources to inform their research papers, but no other uses of the Internet in the schooled space were very common. A decade later when I went back to teach at that same high school I attended as a teenager, Internet access had moved out of the library. Classrooms had Internet connectivity via three terminals pushed against the back wall. This physical arrangement was a sort of visual metaphor for the place of technology in the classroom: not central to learning, not for everyone, and bulky and cumbersome. Even if as a teacher there I had wanted to employ the technology in a meaningful way (I had tried, for example, to use it to publish student work, to give students a real audience for their classroom writing assignments), administration would not allow it. At that time, in the early 2000s, the Internet was still novel and there was fear and uncertainty around technology integration in general, but especially with regard to the Internet.

These days, technology integration is federally mandated, but some of the old anxieties around technology in school remain. Teachers assert that kids don't want to learn information anymore since they know they can Google search anything they want using their ever-present and connected mobile devices. The title of Nicholas Carr's (2008) *Atlantic* article summarizes the fear underlying this lamentation: Is Google Making Us Stupid? Indeed, the general public has long been suspicious of the Internet's effect on memory, retention, depth of cognition, and concentration. Incidentally, other technologies such as calculators have been similarly maligned for hurting students' abilities to think abstractly and to perform calculations themselves.

There has been a comparable critique around the Internet's effect on language: that writing in 140-character bursts on Twitter precludes deep engagement. The evidence often cited is that young people are eschewing conventional spellings for text-speak. And instead of taking notes in class, students are just taking pictures of the board. We wonder, are our technology tools killing language?

Such views, though widespread, are reductionist. Rather than take a technological determinist perspective that sees a clear, one-way, causal relationship between technology and social practices or an essentialist perspective, which assumes that technology is something that can be understood as a fixed, monolithic entity with traceable effects, I endorse a social constructivist perspective (Grint & Woolgar, 1997) in the discussion which follows. Technological determinism posits that technologies give rise to new social practices. The essentialist perspective holds that technology has a core set of traits that are fixed and stable. Both of these views distance technologies from the human beings who both bring them into being and take them up and make them meaningful. The social constructivist perspective sees neither the social nor the technological element as primary in the equation; technology is best understood as a social construction and technologies are meaningless outside of their contexts of use and therefore cannot have any essential character or independently drive social change. This perspective is constitutive of a connective ethnographic approach.

What does matter then when we explore the social uses of technology tools? There are some material aspects of technologies that afford, support, or even encourage particular social practices. The next section offers an explanation of the dynamics among the architecture, affordances, and social uses of technologies and explicates some of the defining features of the mobile phone as a composing tool including the camera, the digital keyboard, applications, and its mobility.

Architecture and Affordances

Of course, technologies are not homogeneous. Tools and platforms vary in *architecture*, or the way they are structured, and *affordances*, or what they allow users to accomplish. Some affordances are synchronous, happening in real time, like chat and other highly interactive sites of discursive action. Others are asynchronous, like blogs, usually with a lag between publishing and consumption. Some, like blogs, are highly malleable, and others, like Facebook, are more rigidly structured. Some are open, and others happen privately, like email. As boyd (2011) explains:

> In essence, the architecture of a particular environment matters and the architecture of networked publics is shaped by their affordances. The common dynamics fall out from these affordances and showcase salient issues that participants must regularly contend with when engaging in these environments. (pp. 39–40)

For example, some digital media are designed for use in "real time," often called "synchronous," such as applications that support chat. The architecture and affordances allow for or encourage certain common dynamics when that technology is taken up. Others are designed to be more monologic, like websites, with no real interaction or talking back. Again, the architectural design of the site is what affords only one-way discourse. Some, like blogs, are highly malleable, and others, like Facebook, are more rigidly structured. For example, a randomly-selected sample of 10 blogs might show a wide variance in design, content/subject matter, and media integration. In contrast, a randomly-selected sample of 10 Facebook pages—a site that affords a limited number of conventions, such as status updates, photo tagging, commenting, and photo-album creation, might look very similar. This, too, affects the way the common dynamics "fall out" (boyd, 2011, p. 39). The affordances are often constrained or enabled by the site's architecture.

In online spaces for composing, adolescents must contend with a number of constraints on communicating, but simultaneously they are also afforded a host of resources for meaning-making, including but also beyond the linguistic. In terms of constraints, one must have access to the hardware and connectivity to compose in the online space. Some spaces, like Twitter, limit users to 140 characters. The smartphone is conducive to "microblogging" or blogging using short message "bursts" of 140 characters or fewer. Traditional blogging practice is rare among youth, but teens microblog fervently (Madden, et al., 2013). Some encourage short posts, or text curation (gathering and displaying already-existing digital content), like Tumblr. Some afford the option of integrating images, hyperlinks, or other media like embedded video, whereas others best accommodate communication through linguistic text. Some allow linking and other means of visibility, whereas some allow privacy. And as the world moves rapidly toward the era of Web 3.0, known as the Semantic Web, all of this is subject to change.

Affordances of the Mobile Phone

While in the past owning a mobile phone was a sign of wealth, now, many young people, even those from the working class or living below the poverty line, have mobile phones and rely on them as their primary Internet connection (Duggan & Smith, 2013). In fact, a 2013 study found no significant difference in smartphone ownership between teens from low-income households and the highest-income households surveyed (Smith). This book deals with the literacy practices of three adolescents who lived in a community that was economically depressed and rurally located and thus many local residents relied on their phones for affordable and reliable Internet access.

Increasingly, people are turning to Internet-enabled mobile phones as a more affordable means of Internet access and the "digital divide is smaller for smartphone use than for Internet use across race" (U.S. Census Bureau, 2012). The participants in this study lived in a rural area of the southeastern United States. In 2010, 57% of rural households had broadband Internet access, compared to 72% in urban areas, according to a November 2011 report by the U.S. Department of Commerce. In many rural places, lack of technology use is not primarily a function of social forces or even economic limitations but is due instead to infrastructural restrictions. These are often tied to challenges associated with geography, such as rugged terrain and dense forested land. Recent data on Internet-equipped (smart) phones as a means of lessening the so-called rural Internet gap cite mobile Internet access as a means of dealing with the infrastructural constraints to bringing broadband Internet to rural places.

While they could use the keyboards on their phones to type out alphabetic messages, the affordances of their mobile phone that emerged as central to the literacy practices of these particular youth were the camera and the mobility of the phones. The architecture (or structure) of the apps they used to compose (namely, Twitter and Instagram) were also salient features of their composing tools that seemed to influence their composing practices.

Camera. In addition to the keyboard, which allowed the youth to communicate through linguistic text, they relied heavily on their cellphone cameras to compose. At times they would capture friends, their bodies in a mirror or using a front-facing camera, or arrange objects to tell a story. Their photos represented bricolage (Lévi-Strauss, 1966) in that they used a wide variety of what they could find in the physical spaces they inhabited to communicate. The camera as a means of multimodal composing seems to engender this approach of using what is available to be captured for communication. With their cellphone cameras, they could easily, quickly, and fluidly communicate powerfully through a visual image, oftentimes removing the need for the intermediary of linguistic text to paint a picture.

Mobility. The portability or mobility of the youth's phones allowed for almost persistent access to online spaces for composing. This permitted spontaneous composing and the creation of texts that were often reactive to the conditions of the physical spaces through which the youth moved, rather than planned or motivated by producing some particular end product (cf. Leander & Boldt, 2013). The youth would often post photographs that captured their physical environment at a given moment or their musings in that moment. Because the phone was always there, a thought could become an alphabetic message pushed out to online followers with relative ease. Composing was less of a formalized, premeditated process, and happened throughout the day for these youth, as one participant, Eli, describes:

Julie:	You usually get [online] at home?
Eli:	Yeah, but like um when I'm waiting for someone ... because I'm not at home most of the time but when I am I'm usually on Twitter like when I'm doin' somethin' and I'm bored I'll go on Twitter.
Julie:	Ok, so wait, you said where most of the time?
Eli:	Like going somewhere, like anywhere, just when I'm bored.
Julie:	Oh, going somewhere.
Eli:	Yeah.
Julie:	You mean like in the car?
Eli:	Yeah, like in the car or somethin'.
Julie:	Oh, ok. Yeah, I do that, too. Just if you're bored. So you don't like think about it that much. It's just sort of like somethin' to do if you're bored?
Eli:	Yeah.

Eli's digital composing practice happened in the in-between spaces of his life, when he was bored, waiting for someone, or stuck in the car. It was not a practice to which he devoted much energy, according to his account, but was a way to pass the time. Indeed, many of Eli's postings seemed to have been made in transit, as the geolocation of many of his photos were shown as on highways and other thoroughfares. Oftentimes, Eli would post photos he had taken at an earlier time as he traveled in the car from one activity to another in his daily life. Photos taken at school were geolocated as having been posted on the highway.

According to Manovich (2001), the haptic, or touch-operated feedback systems in screen media allow us to travel through media (i.e., they make it navigable) and as such users can inhabit many spaces at once. As Eli sensed he was in a state of boredom or waiting or the in media space of transit, he could move into a new space wherein he could chat with friends, express himself, or locate a source of entertainment. His location in space was thus related to his digital-literacy practice in that one motivated the other. Mobility was central to the youth's literacy practice in this way as well.

Architecture of Apps. The youth in this study moved from Twitter, where they primarily communicated through short alphabetic texts, to Instagram, a social photosharing site, over the course of the study. These are two platforms that are usually accessed by smartphone applications (software designed for use on mobile phones). The youth in this study described to me the process of composing and publishing texts through these applications and cited the importance of the fact that these apps saved their log-in credentials thereby mitigating or removing an impediment to swift composing and publishing of their online communication.

Twitter is a popular microblogging website where communication happens in short text bursts of 140 characters or fewer called tweets. All participants tweeted, or posted original tweets to their Twitter pages (which in turn, go out to their "followers," those who subscribe to their updates). They also all engaged in the practice of text curation through "retweeting" on Twitter. In this practice of curation, digital writers create a new text (their Twitter feeds and favorites lists) using others' content quickly and fluidly and without permissions or required citation. These curating features are built into the Twitter interface. Twitter designed the retweet feature as a means of "secondary content promotion." Most users retweet content to spread it further, because a retweeted message goes out to all those who subscribe to one's tweets, their followers.

Near the end of the study, participants' engagement on Twitter had drastically slowed in favor of interaction on a site called Instagram. Instagram is a social photography site where users can post photos, videos, and short messages, often in concert with hashtags, to label and organize photos as well as to add photos and videos to "conversations" around hashtags. On Instagram, a user can "follow" another user, subscribing to that user's posts. Instagram posts consist of a photograph along with a caption and often hashtags, words or phrases preceded by a pound symbol (#) that label the posting and allow it to become searchable by that keyword. Some users also use hashtags as a sort of metacaption (and in some cases, they are used as both a label and a metacaption at once).

Encouraged Practices

There are some practices that seemed to be encouraged by the presence of the particular affordances and architecture of the mobile phone, namely digital curation and social photosharing. For example, on phones with small or virtual keyboards, it was cumbersome if not impossible to create long alphabetic or complex multimodal compositions, but easy to curate texts, because that could be accomplished with one click. The less friction involved in engaging in a particular practice with the mobile phone the more likely it would be to be thought of as a practice encouraged by the mobile phone, in other words. Thus, it is safe to conclude that youth who engage in digital composing primarily by way of mobile phones might approach composition in different ways from those who compose with a computer because the tools encourage different kinds of practices because of their attendant affordances and architectures. However, even these practices are subject to interpretation, and throughout the course of this research I observed a wide variance even within the practices of digital curation and social photosharing.

Digital Curation. Because of the major communicative affordances of the phone, including the camera, the cumbersome digital keyboard, and the architecture of the apps such as the retweet and favoriting features, the participants relied heavily on digital curation as a means of composing. It offered less friction and was a means of communicating powerfully in nearly an instant. Digital curation involves selecting, compiling, and displaying existing digital content rather than creating from the "ground up." In other words, typing long texts or creating complex multimodal texts on the phone was difficult, but curation could be accomplished with the touch of a button. However, curation is a form of composing that is easy to engage in with a mobile phone because there is so much semiotic richness that can be accessed with just the click of a (retweet) button. In curation practice, the practices of reading and composing are nearly collapsed. The youth had to first read their feeds in order to "grab" a tweet or text to integrate into their own. This laminated (Bakhtin, 1982; Prior, 1998) the composing process in that the reading and writing were layered and almost seamless. In the practice of curation, there was no "planning" in the conventional sense; instead, the process was a continuous, fluid one, one in which reader and writer collided, the boundaries between author and audience blurring.

Social Photosharing. Phone-based access shapes literacy practice because of the digital keyboard, which makes the typing of long alphabetic texts cumbersome, as well as the affordance of the camera for communicating with images, an increasingly popular communicative practice with 52% of Internet users posting photos online (Duggan, 2013). In addition, the architecture of the sites the participants in this study used supported the sharing of photographs. In their use of the cameras within their mobile phones, the participants captured (and sometimes stylized and then captured) their physical worlds to communicate ideas without encoding ideas in alphabetic text. They used their faces, bodies, objects, and physical worlds to communicate through selfies and other digital photography. The use of the built-in camera (and apps that did not require logging in or other interruptions to fluidity of process) as composing tools meant that the participants often composed rapidly and spontaneously. This was not always the case, however; sometimes participants put time and energy into stylizing their photographs. In either case, their texts were in dialogue with their physical worlds and demonstrated an immediacy of expression that cut out the middleman of linguistic text to directly illustrate an idea.

Research on Composing

Throughout this book, I refer to the process of creating digital texts as "composition." Others have used "designing," (Edwards-Groves, 2011; Jewitt, 2008; New London Group, 1996), "producing," (Walsh, 2009), "creating," (Zammit, 2007) and "creative media writing" (Rantala & Korhonen, 2008), among other terms. Perhaps my penchant for the term composition comes primarily from my background as an English teacher, but I maintain that composition is a fitting term that is used to describe composing with multiple modes in the fields of music and art, which Manovich (2001) aligned with digital multimodal communication.

Over the past decade, researchers have carried out rigorous and illuminating research in the area of youth digital literacy practice. However, much of this research focuses on the engagement of exceptional youth. Adolescent girls, the demographic who blog the most, have been studied extensively (e.g., Davis, 2010; Guzzetti & Gamboa, 2005), although Technorati found that 95% of blogs are abandoned, so long-term bloggers are rare (Quenqua, 2009). Digital storytelling has been widely studied even though it is a digital composing practice uncommon among youth (e.g., Hull & Nelson, 2005; Lundby, 2008). Other studies of exceptional youths' digital engagement include English language learners using the web to practice English (e.g., Black, 2007, 2008, 2009; Lam, 2000, 2004, 2006, 2009; Lam & Rosario-Ramos, 2009; Lam & Warriner, 2012), youth composers of fanfiction (a niche subculture) (e.g., Black, 2009; Chandler-Olcott & Mahar, 2003; Jenkins, 1994; Mackey & McClay, 2008), and teens interacting in virtual worlds (e.g., Thomas, 2004, 2005, 2007), an activity taken up by just 8% of teens (Lenhart, Purcell, Smith, & Zikhuhr, 2010). While this work has provided rich insight into the digitally literate lives of some adolescents, there is little work that highlights the naturally occurring digital literacy practices of mainstream youth.

It is becoming increasingly important to recognize and examine the digital literacy practices of mainstream youth, in addition to exceptional youth, as Internet access expands to include most young people. For the past two decades, the public has debated the impact of the "digital divide"—the gulf between those who have access to the tools and connectivity to participate in the so-called post-industrial society and those who do not. Experts on the digital divide have criticized the construct for presenting a false binary (e.g., Warschauer, 2009); others have pointed to "second-level divides," or differences in not just access alone but how people take up and use technologies as salient (e.g., Hargittai, 2002).

Literacy research is beginning to take a "spatial turn" (Mills & Comber, 2013) as digital media challenge previously dominant conceptions of space as a container. Today, people move not only with (in the case of mobile phones, for instance) but also through (as they navigate the spaces of) new media. According to theories of space as socially produced (e.g., Lefebvre, 1991; Soja, 1989), instead of being tied to one physical place, ideas, markers of identities, and relationships create new spaces as they travel across the physical or geographic boundaries of the material world. From this perspective, digital literacies are traveling practices (Clifford, 1992) and are, at the same time, practices that produce space. Framing space in this way can help move the typical conversation about the allocation of technological resources in rural places to examine other salient aspects of space like the social use of such tools, for example. Moving beyond the digital divide binary, this chapter explores the spatial qualities of the composing practices of the participants as they composed with their mobile phones across spaces.

The analysis in this book is located at the nexus of digital divide research and digital literacy research. Both bodies of work are relatively new, and, on the whole, rely on different approaches: the former, quantitative studies counting access, and the latter, qualitative studies offering descriptions of people's use of digital tools in communication. In addition, this study is part of an emerging body of research on space and place in education, sometimes called "the spatial turn" (Mills & Comber, 2013) that seeks to address the inequitable distribution of resources, connections, and other capital by taking on a spatial lens. However, in spatially-focused research, the material world (often thought of as place) is only one part of the "field of relations" or "nexus of practice" (Olwig & Hastrup, 1997, p. 8). Spatially-focused research also acknowledges that space is in part conceived or imagined (Lefebvre, 1991; Soja, 1989).

Digital Divide

This study took place in a small, rurally-located town. When it comes to rural places and the construct of the digital divide, there is much confusion. For example, conflation of the descriptors "rural" and "low-income" lead to misconceptions; in addition, lack of attention to age as a variable in discussions of technology access and ownership has muddied the waters (on the whole, more young than old take up technologies, and on the whole, more old than young people live in rural places). Much work on the digital divide has focused on general trends; researchers have examined correlating variables (i.e., sociodemographic characteristics) in various settings over the past twenty years since concerns about access gaps were originally raised (e.g. Haight, Quan-Haase, & Corbett, 2014; Ritzhaupt, Feng, Dawson, & Barron, 2013). Quantitative studies that measure access offer a rough assessment, but rich pictures of technology in use in rural areas are key to understanding the nature of the so-called digital divide.

Digital Literacy Practices

Education researchers have carried out robust and edifying research in the area of youth digital literacy practice over the past decade, but most this research focuses on the engagement of urban youth (e.g., Black, 2009; Chandler-Olcott and Mahar, 2003; Davis, 2010; Guzzetti & Gamboa, 2005; Hull & Nelson, 2005; Jenkins, 1992; Mackey and McClay, 2008). Such work spoke back to the once popular notion that digital media engagement was corrupting the English language (e.g., Helderman, 2003; Lee, 2002), was frivolous in nature (e.g., Rainie, 2011), or was stupidity-inducing (e.g., Carr, 2008). By illuminating all the ways youth demonstrated literacy in their lives, these scholars responded also to the assertion that youth were falling behind the rest of the world in the area of literacy post-*A Nation at Risk* (Lankshear & Knobel, 2008). However, there is a dearth of work that highlights the naturally-occurring digital literacy practices of youth from rural communities. Moving beyond the digital divide binary, such studies can help literacy educators support the schooled literacy development of those students in rural communities who are creative and dexterous users of technology and offer insight into rich and meaningful uses of technologies by people in rural places.

Though limited, work in this area has afforded a multifaceted picture of the use of digital media and technological tools by rurally-located youth in varied places. Patnak-Shelat and DeShano (2014) examined the role of digital media in the "lifeworlds" of youth in rural India, tracing the intersections of the dominant discourses of place, the daily social lives of youth, and digital technologies. The researchers found that despite technology's affordance of mobility, youth used digital media to accomplish personal goals related to their immediate material place and their social lives in those places. This is an important counter perspective to the overly propitious view of technology as panacea for all that ails youth in rural places.

This is not to say that rural youth have failed to harness the transformative potential of technologies to remake social spaces. Gray (2009) studied the use of digital media by rurally-located queer youth and found that youth created new spaces and went beyond the immediate boundaries of the physical spaces of their lives to explore their identities and connect with others. Gibbons (2014) continued a line of inquiry focused on defining rural media literacy (e.g., Gibbons, 2010) through her work with rural filmmakers. Considering youth media production as a way of representing self, place, and social space, this work examines the meaning of place in digital literacy practice as well as how counternarratives that challenge stereotypes are constructed in youth media production. In this work, the focus is shifted from the transformative potential of available tools to how such tools are exploited by youth to tell and to share their stories across time and space.

It is becoming increasingly important to recognize and examine the digital literacy practices of youth other than just the urban and exceptional youth often studied as access expands. In particular, since phone-based in distinct from other kinds of digital or Internet-based literacy practices because of the particular affordances and constraints of these mobile tools, a base of knowledge of these practices can help stakeholders to provide more and richer opportunities to learn for students who use these tools in the wake of a decades-long debate around the implications of the so-called digital divide.

Such research is important too because it challenges notions of rural youth as disconnected and anachronistic in a digital age. The participants in this research used their phones to achieve a variety of life goals including both moment-to-moment goals and larger life goals like identify formation. They communicated in short texts and via photographs and through practices like favoriting in ways that were reflective of Internet culture at large. They harnessed the ethos of Web 2.0 to interact with real audiences for real purposes. In some ways, too, the participants adopted practices and conventions afforded by the flows of the global interactive space. At the same time, however, their scaled interactive spaces were largely reflective of connections from their material words. In other words, for the most part, the youth in this study used their access to the Internet in ways that were in line with their offline experiences of the world.

Is the goal for students in the 21st century to simply have access to technology tools or to become "cosmopolitan" and have an amplified global voice? This will be a question that schools will soon have to grapple with. Although the youth in this study were afforded iPads to use at school and owned Internet-equipped phones, they lacked further support (e.g., wireless Internet connections to use the iPads at home, instruction about the global space) to harness the transformative potential of the tools. The iPads became little more than aluminum bricks when the participants got home and did not have wireless Internet connections with which to take advantage of the iPad's features.

The one-to-one iPad program at White Birch was implemented as a means of countering the effects of poverty in the area on education through access to cutting edge technology. This kind of understanding of social justice is built on the "distributive paradigm" (Young, 1990) that oversimplifies justice by equating it to the simple presence of absence of economic resources without questioning the propriety of particular resources in a specific context. Roberts and Green (2013) criticize such "universalizing notions of justice and a lack of recognition" (p. 768). One-to-one iPad programs like this one seem to signal the arrival of a new factory model of schooling for the global marketplace, given that there were few, if any, high-tech jobs in Brookeville for which exposure to iPads would prepare youth.

Metronormative Policy

Technology integration is accounted for within the Common Core State Standards and the newly-passed Every Student Success Act (ESSA) authorizes federal funding for a number of STEM priorities. The implications of federally-sanctioned technology use are twofold. First, the "geographical blindness" (Green & Letts, 2007) of neoliberal policy that results from the inherent "metronormativity" (Green, 2013) can prove problematic for rural social spaces. In policy, rural places exist as "quasi-urban social constructions" (Roberts & Green, 2013, p. 766) and the resulting blanket response to the monolithic construct of educational need equates the needs of urban and diverse rural sites (Roberts & Green, 2013). The implementation of large-scale technology programs like the one at White Birch High School is one such example. Certainly, as technology continues to expand and evolve, stakeholders seek to understand how to prepare students for digitally saturated futures while meeting the demands of assessment-centered contexts for literacy learning. This is not to say that proponents of place-based instruction see technology as irrelevant to rural places; they are also, however, interested in honoring the unique goals of students from rural communities (Azano, 2014) and in considering the reasons for technology integration.

The second issue raised by federal technology mandates is the spirit behind the imperative—that of global competitiveness. Rural schools often struggle to reconcile federal mandates centered on producing globalized members of the "world's most competent workforce" (Ochshorn, 2011) and the unique economic, political, and social circumstances in their local communities. If rural literacies are, as Donehower, Hogg, and Schell (2007) suggest, "literate skills needed to achieve the goals of sustaining life in rural areas" (p. 4), more inquiry into the ways that technology can be taken up to support the goals and sustain the lives of people in rural areas needs to be performed.

Policy Opportunities

Recent years have seen a specific policy focus on extending opportunity to underserved communities and groups intended to support their success in an increasingly globalized and technology-focused economy. This policy focus has materialized in the form of increased funding for access to high-speed Internet, support for courses and curriculum that center on science, technology, engineering and math (STEM) including computer programming or coding, and funding for teacher training to help teachers to integrate technology into their teaching and support student learning in STEM subjects. With the recent passing of the Every Student Succeeds Act (ESSA) which amended the Elementary and Secondary Education Act (ESEA), there promises to be an increased focus on STEM education in particular as implementation of the new law happens in the coming years.

Since President Obama took office in 2008 as President of the United States and over the course of his administration, there was increased access

to technology and connectivity in schools. In particular, since he launched his ConnectED initiative in 2013, which aimed to provide next generation Broadband Internet access to 99% of schools in the United States, the connectivity divide in schools has been cut in half. In 2013, only 30% of school districts were meeting the President's goal of 100 kilobits per student, whereas in 2015, 77% of districts were meeting that goal. ConnectED partners have provided software, devices, and digital tools like educational applications to five million students as of 2016.

In addition to access to high-speed Internet, federal and local supports for curricula focusing on technology integration is a policy imperative when we talk about providing students with access to 21st century knowledge and skills. Typically, we have observed disparities in demographic groups who have access to STEM courses, particularly computer science courses. For example, in 2015, fewer than 15% of all high schools offered any Advanced Placement (AP) computer science courses and only 22% of those who took the exam were female and only 13% were African American or Latino. There were three states in which no girls took the AP computer science exam at all and in 10 states, fewer than 10 girls took the exam statewide. There were 21 states in which fewer than 10 African American students took the AP computer science exam and in nine of those states, no students of color took the exam at all. These inequalities in access to STEM-related coursework are striking when we consider that by 2018, it is projected that 51% of all STEM jobs will exist in computer science-related fields. In response, President Obama introduced Computer Science for All, with his budget calling for four billion dollars in state funding with one hundred million dollars going to districts for K-12 computer science training for teachers as well as instructional materials. He also called for additional funding for computer science education through the National Science Foundation and the Corporation for National and Community Service. While as of 2015, computer science was absent from P-12 education in most districts, with a mere twenty-seven states counting computer science courses toward graduation, this increased funding for computer science should help to train a sufficient amount of teachers to teach computer science courses (thought to be central to this dearth of computer science course offerings) and to change the perception of the importance of computer science curriculum at the State level.

The lack of teacher workforce preparation for teaching computer science and other STEM-related courses is also being addressed at the Federal level. As he left office in 2017, President Obama was halfway to meeting the goal that he set in 2011 of preparing 100,000 new math and science teachers by 2021 because of public-private partnerships. An example of one such initiative is the partnership between electrical wire and cable manufacturer Southwire and Carroll County Schools in Georgia. This initiative has expanded access to STEM-related learning in real-world contexts for at-risk students through i3 grant funds. Several other technology, education, philanthropic, and corporate partners have pledged to support the integration of computer science in communities all across the United States including STEM-teacher education. In the Every Student Succeeds Act, STEM programs are an allowable use of funds under Title IV, and the STEM Master Teacher Corps provision is an optional reservation of funds within Title II National Activities. The latter means that the Education Secretary may establish a competitive grant program for states to develop a STEM Master Teacher Corps, which would amount to a dedicated funding stream for s STEM Master Teacher Corps.

Conclusion

In the span of just twenty years, schools in the United States have moved from an ethos of fear and a place of scarcity when it came to technology integration to acceptance and celebration. In fact, technology in some places is regarded as a cure-all for what ails low-income and rural schools. Although there are persistent issues related to access (one being the homework gap, when students have school assignments that require Internet access but have no reliable Internet connection in the home), there has been increased support from the Federal level down for the provision of hardware and software, technology-focused curriculum, and teacher training that result in increased opportunity for students to learn with and through technologies. The remaining challenge will be ensuring that rurally-located students, students of color, and female students, all of whom have historically had limited access to technology-infused learning experiences, have opportunities to engage with the architecture and affordances that digital tools offer in ways that support rich literacy learning. Also, with many students accessing the Internet by way of mobile phones, educators and other stakeholders must remain aware that a digital divide still exists in terms of hardware. It can be difficult, clunky, or even impossible to complete some homework assignments requiring Internet access with mobile tools alone.

In addition to these additive imperatives, it is time that mobile phones, which most students bring with them into the classroom every day, be allowed to emerge from their hiding places in backpacks and student pockets. Students can use the capabilities of their mobile phones to access information to achieve moment-to-moment goals; to compose multi-modally with photography, video and linguistic text; to compose collaboratively and inter-textually; and to learn about the world through exploring online spaces and making connections. To eschew the powerful tools that travel with youth throughout their daily lives is at best false and at worst irresponsible as we seek to prepare students to be participants in a future that is increasingly digitally-saturated. The pages that follow explore these rich practices in greater depth.

References

Azano, A. P. (2014, February 14). Glocalized learning: What's relevant in the rural world [Web log post]. Retrieved from http://smartblogs.com/education/2014/02/14/glocalized-learning-whats-relevant-in-the-rural-world/

Bakhtin, M. (1982). *The dialogic imagination.* Austin, TX: University of Texas Press.

Black, R. W. (2007). Fan fiction writing and the construction of space. *E-Learning and Digital Media*, *4*, 384–397.

Black, R. W. (2008). *Adolescents and online fan fiction.* New York, NY: Peter Lang.

Black, R. W. (2009). Online fan fiction, global identities, and imagination. *Research in the Teaching of English*, *43*, 397–425.

boyd, d. (2011). Social network sites as networked publics: Affordances, dynamics, and implications. In Z. Papacharissi (Ed.), *A networked self: Identity, community and culture on social network sites* (pp. 39–58). New York, NY: Routledge.

Carr, N. (2008, July 1). Is google making us stupid? What the internet is doing to our brains. Retrieved from http://www.theatlantic.com/magazine/archive/2008/07/is-google-making-us-stupid/306868/

Chandler-Olcott, K., & Mahar, D. (2003). "Tech-savviness" meets multiliteracies: Exploring adolescent girls' technology-mediated literacy practices. *Reading Research Quarterly*, 38, 356–385. doi:10.1598/RRQ.38.3.3

Clifford, J. (1992). Travelling cultures. In L. Grossberg, C. Nelson, & P. Treichler (Eds.), *Cultural studies* (pp. 96–116). New York, NY: Routledge.

Davis, K. (2010). Coming of age online: The developmental underpinnings of girls' blogs. *Journal of Adolescent Research*, *25*, 145–171. doi:10.1177/0743558409350503

Donehower, K., Hogg, C., & Schell, E. (2007). *Rural literacies.* Carbondale, IL: Southern Illinois University Press.

Duggan, M. (2013, October 28). *Photo and video sharing grow online*, Pew Internet & American Life Project. Retrieved December 12, 2013 from http://www.pewinternet.org/2013/10/28/photo-and-video-sharing-grow-online/.

Duggan, M., & Smith, A. (2013, September 16). *Cell internet use 2013*, Pew Internet & American Life Project. Retrieved October 12, 2013 from http://www.pewinternet.org/Reports/2013/Cell-Internet.aspx.

Edwards-Groves, C. (2011). The multimodal writing process: Changing practices in contemporary classrooms. *Language and Education*, *25*, 49–64. doi:10.1080/09500782.2010.523468

Gibbons, D. (2010). Tracing the paths of moving artifacts in youth media production. *English Teaching: Practice and Critique*, 9(1), 8–21.

Gibbons, D. (2014). Rural places meet media literacy: Representing truth and self in rural social space. *Journal on Images and Culture* (6). Retrieved from http://vjic.org/vjic2/?page_id=3057

Gray, M. (2009). *Out in the country: Youth, media, and queer visibility in rural America*. New York, NY: NYU Press.

Green, B. (2013). Literacy, rurality, education: A partial mapping. In B. Green & M. Corbett (Eds.), *Rethinking rural literacies: Transnational perspectives* (pp. 17–34). New York, NY: Palgrave Macmillan.

Green, B., & Letts, W. (2007). Space, equity, and rural education: A "trialectial" account. In K. N. Gulson & C. Symes (Eds.), *Spatial theories of education: Policy and geography matters* (pp. 57–76). New York, NY: Routledge.

Grint, K., & Woolgar, S. (1997). *The machine at work: Technology, work, and organization*. Cambridge: Polity.

Guzzetti, B., & Gamboa, M. (2005). Online journaling: The informal writings of two adolescent girls. *Research in the Teaching of English*, *40*, 168–206.

Haight, M., Quan-Haase, A., & Corbett, B. (2014). Revisiting the digital divide in Canada: The impact of demographic factors on access to the Internet, level of online activity, and social networking site usage. *Information, Communication, & Society*, *17*(4), 503–519.

Hargittai, E. (2002). Second-level digital divide: Differences in people's online skills. *First Monday*, *7*(4), Article 2.

Helderman, R. S. (2003, May, 20). Click by click, teens polish writing; instant messaging teaches more than TTYL and ROFL. *The Washington Post*, p. B.01.

Hull, G., & Nelson, M. (2005). Locating the semiotic power of multimodality. *Written Communication*, *22*, 224–261.

Jenkins, H. (1992). *Textual poachers: Television, fans, and participatory culture*. New York, NY: Routledge.

Jenkins, H. (1994). Do you enjoy making the rest of us feel stupid? Alt.tv.twinpeaks, the trickster author and viewer mastery. In D. Lavery (Ed.), *Full of secrets: Critical approaches to Twin Peaks* (pp. 51–69). Detroit, MI: Wayne State University Press.

Jewitt, C. (2008). Multimodality and literacy in school classrooms. *Review of Research in Education*, *32*, 241–267.

Lam, W. S. E. (2000). Second language literacy and the design of the self: A case study of a teenager writing on the Internet. *TESOL Quarterly*, *34*, 457–482. doi:10.2307/3587739

Lam, W. S. E. (2004). Second language socialization in a bilingual chat room: Global and local considerations. *Language, Learning and Technology*, 8(3), 44–65.

Lam, W. S. E. (2006). Culture and learning in the context of globalization: Research directions. *Review of Research in Education*, 30, 213–237. doi:10.3102/0091732X030001213

Lam, W. S. E. (2009). Multiliteracies on instant messaging in negotiating local, translocal, and transnational affiliations: A case of an adolescent immigrant. *Reading Research Quarterly*, 44, 377–397. doi:10.1598/RRQ.44.4.5

Lam, W. S. E., & Rosario-Ramos, E. (2009). Multilingual literacies in transnational digitally-mediated contexts: An exploratory study of immigrant teens in the United States. *Language and Education*, 23, 171–190. doi:10.1080/09500780802152929

Lam, W. S. E., & Warriner, D. (2012). Transnationalism and literacy: Investigating the mobility of people, languages, texts, and practices in contexts of migration. *Reading Research Quarterly*, 47, 191–215. doi:10.1002/RRQ.016

Lankshear, C., & Knobel, M. (Eds.). (2008). *Digital literacies: Concepts, policies, and practices*. New York, NY: Peter Lang.

Leander, K., & Boldt, G. (2013). Rereading "a pedagogy of multiliteracies": bodies, texts, and emergence. *Journal of Literacy Research*, 45, 22–46. doi:10.1177/1086296X12468587

Lee, C. D. (2002). Interrogating race and ethnicity as constructs in the examination of cultural processes in developmental research. *Human Development*, 45(4), 282–290.

Lefebvre, H. (1991). *The social production of space*. Cambridge, MA: Blackwell.

Lenhart, A., Purcell, K., Smith, A., & Zikhuhr, K. (2010, February 3). *Social Media and Young Adults, Pt. 3: Social Media*. Retrieved from http://www.pewinternet.org/2010/02/03/part-3-social-media/#adults-teens-and-virtual-worlds

Lévi-Strauss, C. (1966). *The savage mind*. Chicago, IL: University of Chicago Press.

Lundby, K. (Ed.). (2008). *Digital storytelling, mediatized stories: Self representations in new media*. New York, NY: Peter Lang.

Mackey, M., & McClay, J. K. (2008). Pirates and poachers: Fan fiction and the conventions of reading and writing. *English in Education*, 42(2), 131–147.

Madden, M., Lenhart, A., Cortesi, S., Gasser, U., Duggan, M., Smith, A., & Beaton, M. (2013). *Teens, social media, and privacy*. Retrieved from http://www.pewinternet.org/Reports/2013/Teens-Social-Media-And-Privacy.aspx

Manovich, L. (2001). *The language of new media*. Cambridge, MA: MIT Press.

Mills, K. A., & Comber, B. (2013). Space, place and power: The spatial turn in literacy research. In K. Hall, T. Cremin, B. Comber, & L. Moll (Eds.), *International handbook of research in children's literacy, learning and culture* (pp. 412–423). London: Wiley-Blackwell.

New London Group. (1996). A pedagogy of multiliteracies: Designing social futures. *Harvard Educational Review*, 66, 60–92.

Ochshorn, S. (2011). Forging a new framework for professional development. *A report on The science of professional development in early childhood education: A national summit*. Washington, DC: Zero to Three.

Olwig, K. F., & Hastrup, K. (Eds.). (1997). *Siting culture*. New York, NY: Routledge.

Patnak-Shelat, M., & DeShano, C. (2014). Digital youth cultures in small town and rural Gujarat, India. *New Media and Society*, 16(6), 983–1001.

Prior, P. (1998). Writing/disciplinarity: A sociohistoric account of literate activity in the academy. Mahwah, NJ: Lawrence Erlbaum Associates.

Quenqua, D. (2009, June 5). Blogs falling in an empty forest. *New York Times*. Retrieved from http://www.nytimes.com/2009/06/07/fashion/07blogs.html?_r=0

Rainie, L. (2011, December 2). *The internet as a diversion and destination*. Retrieved from http://pewinternet.org/Reports/2011/Internet-as-diversion.aspx

Rantala, L., & Korhonen, V. (2008). New literacies as a challenge for traditional knowledge conceptions in school: A case study from fifth graders digital media production. *Studies in Media & Information Literacy Education*, 8(2), 1–15. doi:10.3138/sim.8.2.002

Ritzhaupt, A., Feng, L., Dawson, K., & Barron, A. (2013). Differences in student information and communication technology literacy based on socio-economic status, ethnicity, and gender: Evidence of a digital divide in Florida schools. *Journal of Research on Technology in Education*, 45(4), 291–307.

Roberts, P., & Green, B. (2013). Researching rural places: On social justice and rural education. *Qualitative Inquiry*, *19*(10), 765–774.

Smith, A. (2013). *Technology adoption by lower income populations*. Retrieved from http://www.pewinternet.org/Presentations/2013/Oct/Technology-Adoption-by-Lower-Income-Populations.aspx

Soja, E. (1989). *Postmodern geographies: The reassertion of space in critical social theory*. London: Verso.

Thomas, A. (2004). Digital literacies of the cybergirl. *E-Learning and Digital Media*, *1*, 358–382. doi:10.2304/elea.2004.1.3.3

Thomas, A. (2005). Children online: Learning in a virtual community of practice. *E-Learning and Digital Media*, *2*, 27–38.

Thomas, A. (2007). *Youth online: Identity and literacy in the digital age*. New York, NY: Peter Lang.

U. S. Census Bureau. (2012). *Computer use infographic*. Retrieved December 17, 2014 from http://www.census.gov/hhes/computer/files/2012/Computer_Use_Infographic_FINAL.pdf

Walsh, M. (2009). In pedagogic potentials of multimodal literacy. In L. Tan Wee Hin & R. Subramanian (Eds.), *Handbook of research on new media literacy at the K–12 level: Issues and challenges* (pp. 32–47). Hershey, PA: IGI Global.

Warschauer, M. (2009). Learning to write in the laptop classroom. *Writing & Pedagogy*, *1*(1), 101–112.

Young, I. M. (1990). *Justice and the politics of difference*. Princeton, NJ: Princeton University Press.

Zammit, K. (2007). Popular culture in the classroom: Interpreting and creating multimodal texts. In R. Whittaker, M. O'Donnell, & A. McCabe (Eds.), *Advances in language and education* (pp. 60–76). London: Continuum.

· 3 ·

MOBILE PHONE COMPOSING AND THE SPATIAL TURN

Abstract
This chapter explores the ways in which digital media challenge previously dominant conceptions of space and place and how these new conceptions bear out in youth digital literacy practice. Mobile technologies have provided new sites for examining place, space, and spatiality, and youth digital composing practices, which happen across both physical and digital spaces, illustrate the new ways of thinking about these concepts and for ethnographic research. From the way we think about the nature of space to how we conceive of where spaces originate to how movement across or within spaces happens with the help of mobile technologies, youth mobile phone-based composing practices represent a generative set of habits and dispositions that can act as a starting place for thinking about what funds of knowledge frequent mobile phone composers are bringing into classrooms after "the spatial turn." The chapter ends with a discussion of embodied literacy practices and how the concept of materiality relates to mobile phone-based literacy practices.

Keywords: Space, spatiality, youth digital literacy, mobile technologies, spatial turn, embodied literacy practices, materiality, ethnography, culture, boundaries

This chapter explores the ways in which digital media challenge previously dominant conceptions of space and place and how these new conceptions bear out in youth digital literacy practice. Mobile technologies have provided new sites for examining place, space, and spatiality, and youth digital composing practices, which happen across both physical and digital spaces, illustrate the new ways of thinking about these concepts. From the way we think about the nature of space to how we conceive of where spaces originate to how movement across or within spaces happens with the help of mobile technologies, youth mobile phone-based composing practices represent a generative set of habits and dispositions that can act as a starting place for thinking about what funds of knowledge frequent mobile phone composers are bringing into classrooms.

People move not only with (in the case of mobile phones, for instance) but also through (as they navigate the spaces of) new media. As such, digital literacies are traveling practices (Clifford, 1992). Instead of being tied to one physical place, ideas, and markers of identities and relationships travel as well. They create new spaces as they travel across the physical or geographic boundaries of the material world. As such, digital literacy practices are practices that produce space. Lefebvre (1991) described such practices as social practices that produce social spaces, as meaning-making processes that create real or imagined geographies.

The way we conceptualize digital space has been defined in part as a function of the language people use to describe it and in part by its architecture and affordances. Terms like "cyber*space*," "*home*page," "chat*room*," "email *address*," web*site*," "navigating," "*World Wide* Web," and URL (universal resource *locator*), to name a few, evoke a sense of place, of geography (cf. Manovich, 2001). These terms make it seem like the Internet is a destination to which you travel, some "cyberian apartness" (Miller & Slater, 2000, p. 5). But despite how we might discursively position the Internet and its vestiges, it is not some separate place, but rather it is rooted within people's existing spheres of social practice. Beyond conceptualizations that are rooted in language, though, the architecture and affordances of digital media affect how we think about digital space. For example, since digital media allow us to, as some argue, inhabit several spaces at once, and even move across several physical and digital spaces at once. Digital technologies allow data to travel in ways that facilitate interaction across broad spaces in real time, which provide the illusion of a "flattened" physical world (Friedman, 2005). Readers of networked digital media have the ability to quickly and easily interact with authors and other users through comment boxes, social-networking buttons, or email, making digital composition more potentially dialogical than print composition. In this way, they allow for the impression that seemingly intraversable geographic spaces have been collapsed. In other words, they allow people to organize space differently conceptually. That is to say, technology facilitates a certain kind of movement that has implications for how we think about space.

To further complicate this set of ideas about technology and space and place, technologies have made it possible to produce spaces socially. One can now create a website and upload it to the Web, where it has the potential to become a site of interaction. For example, when a person posts a question to an online message board, he or she creates a space for conversation that can be navigated to and visited by others. At once, these digital texts are texts as well as spaces; they are not just artifacts, but also meeting places where people can share ideas or otherwise communicate. In other words, through the use of technologies, people now meet and dialogue in textual spaces (Davies, 2011; Davies & Merchant, 2009). When youth create online profiles on social networking sites that contain photo galleries, status updates, and postings from friends, they create a space, the nexus of their social connections, which reflects a dynamic construction of identity and social reality.

Relatedly, some youth, shut out of the sanctioned spaces for literacy in the school as "failing" students, engage online in an attempt to create spaces for themselves in which they are able to demonstrate literacy and assert themselves as competent creators and consumers of a variety of texts. In such cases, space is characterized by intersections of power and enactments of identity (Moje, 2004). In addition, as constructed, not fixed, entities, spaces can be contested and reconfigured (Mills & Comber, 2013).

After an exploration of background and recent research in the area of space and place in digital literacy, this chapter presents some accounts of the ways youth move through as well as create spaces through digital literacy practices. It also explores the ways in which youth contest and remake spaces both digital and physical through their engagement with digital literacy practices.

Spaces for Learning: The In- and Out-of-School Dichotomy

Much of the literature on youth literacy practices is place-based. Dichotomizing literacy into "in school" and "out-of-school" literacies places the emphasis on the physical spaces. Bifurcating youth literacies as at-home, unsanctioned literacy, nonacademic, and informal in opposition to at-school, sanctioned, academic, and formal creates several problems. First, this lack of continuity among youth's literacy practices across social spaces causes us to view adolescents' literacy lives as fragmented. These spaces and practices are actually more fluid, with continuity between what occurs at home and what happens at school, among other spaces young people inhabit, because they are not bound. Research on literacy in out-of-school settings has included print literacies, digital literacies, and even dress (Moje, 2000). In addition, constructing such a binary creates a sort of myopia where observers fail to acknowledge school-like practices outside of school or unsanctioned practices in school (Hull & Schultz, 2001, 2002).

Recent literature has examined literacy practice from a spatial perspective that underscores at the mobility of youth practices across spaces, recognizing youth literacies as more continuous. For example, Schmier (2010) explored the ways that literacies from school, a physical place, traveled home, and literacies from home, traveled to school. Schmier's participants brought literacies engaged on social networking platform MySpace into their homework and school homework practices onto MySpace. In this way, the spaces of school, home, and the online space were demonstrated to be inextricable and fluid and a new kind of hybrid space opened up.

Another study that took up the idea of space as socially produced and identities as mobile was Moje's (2000) study with five gang-connected youth. Moje (2000) presented data from three years of research to illustrate how they used their literacy practices "to be part of the story" or to claim a space, construct an identity, and take a social position in their worlds. This study framed literacy as a tool, which Moje characterized as "communicative, expressive, and transformative" (Moje, 2000, p. 653). By viewing literacy as a tool for "exploring possible worlds, claiming a space" (Moje, 2000, p. 654), Moje framed literacy in terms of social practice across spaces to accomplish life goals. In other words, Moje framed literacy in terms of social practice that produced space. Similar studies theorize space in ways similar to agency, as in the space people create for themselves to participate (Moje, 2000; O'Brien, 1998, 2003; O'Brien, Springs, & Stith, 2001).

Lam (2000) conducted a study of a Chinese immigrant student who created his own website and began interacting with online pen pals to improve his English after being assigned to remedial courses at school because of his lack of facility with the English language. Lam (2000) pointed out the way that her participant developed an identity as a skilled user of the English language. Whereas in the schooled literacy space the Chinese teen, Almon, was not seen as a competent English communicator, in the online space, he was able to create a space for himself as a competent English speaker.

While classroom English appeared to contribute to Almon's sense of exclusion or marginalization (his inability to speak like a native) which paradoxically contradicts the school's mandate to prepare students for the workplace and civic involvement, the English he controlled on the Internet enabled him to develop a sense of belonging and connectedness to a global English-speaking community (Lam, 2000, p. 476). Almon contested his identity in the schooled space and created a new identity through online interaction that had real implications on his literacy learning in school. After six months of interacting online, the researcher noted a marked difference in her participant's in-school English performance.

Lam's study has implications for thinking about the intersections of identity and space but also demonstrated the burgeoning need to think beyond place-based notions of literacy and to think instead of social practices across spaces in an era of transnational communication characterized by "global ethnoscapes" (Appadurai, 1991). As the Internet, a global network of interconnected computers or other devices, continues to connect people across traditional borders, it will be vital to have accounts of people's practices, perspectives, and cultural models around writing in the scalable Internet context. Connective ethnography, which allows for a research focus that transcends a specific focus on a boundaried culture or subculture, can be fruitfully used to study such practices in the scalable Internet context.

Research Methodologies and Space

Although new views of space predate "networked society" (Castells, 1996), digital literacies research has necessitated the re-theorization of space. The so-called "social turn" in literacy studies was accompanied by research that situated texts and composers in contexts of practice; as ethnographic research began to consider digital literacies, issues of space and place complicated the notion of "the field" and situatedness (Baron, 2008; Buckingham, 2008; Ito, Okabe, & Matsuda, 2005; Ling, 2004; Livingstone, 2008; Mazzarella, 2005; Stornaiuolo, Hull, & Nelson, 2009).

Researchers are transcending the "container-oriented" approach to research to examine ways people create and use texts across spaces (e.g., Hine, 2000; Jones, 2005; Leander & Lovvorn, 2006; Lemke, 2006). Jones (2001), for example, examined ways youth in Hong Kong moved across virtual and material spaces through their use of instant-messaging software. As they communicated across geographic boundaries while situated locally in a material place, they thus inhabited several spaces simultaneously. Some researchers focused on globalization, diasporas, and transnational communication (Brouwer, 2006; Lam, 2000, 2004, 2006, 2009; Skop & Adams, 2009) and the ways youth engage across a multitude of boundaries and borders. Such views of space are salient now, as the online adds a new layer to (or new nodes within) social space.

The so-called "social turn" in literacy studies was accompanied by research that situated texts and composers in contexts of practice; as ethnographic researchers began to consider digital literacies, issues of space and place complicated the notion of "the field" and situatedness (Baron, 2008; Buckingham, 2008; Ito, Okabe, & Matsuda, 2005; Ling, 2004; Livingstone, 2008; Mazzarella, 2005; Stornaiuolo, Hull, & Nelson, 2009). The work that has troubled existing notions of context in literacies research has been called "the spatial turn" (Mills & Comber, 2013).

Early ethnographic studies of technologies grappled with the conceptualization of the field. Specifically, some researchers dichotomized the online and offline as separate spaces, whereas some saw them as a fluid whole. Turkle's (1984, 1995, 2011) large-scale ethnographic studies explored the reflexive relationships of people and technologies in everyday life. Though carrying out participant observation and interviews online, Turkle also observed in person and made the decision to report if a participant only took part in an interview. Turkle (1995) characterized this choice as a "real life bias" (p. 324) and argued that this approach allows the researcher to present findings with more richness and attention to the complexity of social life.

As technologies have become ubiquitous and more part of people's everyday practices, a concomitant conceptual shift contends with the online–offline dichotomy (evident in terms such as "cyberspace") and views of identity as fragmented. In spite of this, severe limitations to ethnographic research frame Internet practices as cultural and contextually rooted.

Place-based studies that follow participants through on and offline realms lend credence to the view of space as fluid. Miller and Slater (2000) conducted an ethnographic study of Trini people in the online and offline realms. Findings overlapped communicative practices online and offline among Trini people. That is to say, their research showed the ways in which Trini people used the Internet to do the things they had set out to do in their everyday life. They did not go online and escape into a world apart; they go online and practice "being Trini." As the authors argue:

> We need to treat Internet media as in continuum with and embedded in other social spaces … They happen within mundane social structures and relations that they may transform but that they cannot escape into a self-enclosed cyberian apartness. (Miller & Slater, 2000, p. 5)

The authors emphasized that the offline space is not a separate realm but its use is related to identity and local context. Unlike earlier studies, they conceptualized identities as more fluid, viewing the Internet as embedded in people's lives and only taking significance in the ways people use it. Indeed, offline and online are not discrete places, but simply part of people's lives.

Ethnographic studies of young people's digital literacy practices (e.g., Baron, 2008; Buckingham, 2008; Ito, Okabe, & Matsuda, 2005; Ling, 2004; Livingstone, 2008; Mazzarella, 2005) must begin by defining "the field" and context or otherwise theorizing space, as necessary. Connective ethnography (Hine, 2000; Leander & McKim, 2003) is employed by researchers who seek to describe and interpret informants' practices online and offline (boyd, 2008; Hine, 2000; Markham, 1998) through the contexts of school, home, and Web spheres. Ethnography conceptualized in this way becomes a "process of following connections" (Olwig & Harstrup, 1997). This design allows for the recognition of the fluidity of the "real" and "virtual." Connective ethnography makes no distinction between the online and the offline in the nature of reality. Researchers have attested to the "seamlessness" of young people's online and offline realities (boyd, 2008; Thomas, 2007), and some research shows that teenagers' self-presentation and written expression on the Web is in harmony with their offline lives (Blinka & Smahel, 2009; boyd, 2007, 2008; Huffaker & Calvert, 2005; Stern, 2007).

In this study, I utilized this approach by conducting interviews through computer-mediated methods such as electronic messaging, but also in person to reflect the insignificance of the offline–online dichotomy and the importance instead of considering all of the participants' literacy practices in the scope of their sociocultural context. In addition, the offline was often called upon to interpret the online and vice versa.

Typical ethnographic methods of data collection rely on relating social meanings of participants in naturally occurring settings, the participants' territory. I had to contend with the question of how I would define the field in a project that included digital and physical spaces. Early ethnographers defined culture in relationship to geographic place or a group identity tied to a specific place. This approach was logical given that most groups had limited physical mobility at the time (Olwig & Harstrup, 1997). More recently, because people, their tools, and their texts have more potential to travel, ethnographers have come to challenge the notion of place as bounded and identity as tied to geography. Accordingly, Piot (1999), recognizing the cultural flows that have been associated with the term globalization, warns against essentialized notions of peoples and cultures. In his ethnography of the people of Kabre, Togo, he argues that they are also "part of the modern world." He highlights the modern elements of what was popularly considered a "primitive" culture, calling it an "uncontainable hybrid cultural landscape" (Piot, 1999, p. 178).

In the course of conducting this research, for example, I had to grapple with the relationship of place to my inquiry in conducting an ethnographic study that included the online; because of my longstanding history with the Internet, I had some preliminary ideas about this. In the 1990s, I was the only child I knew who had regular Internet access, because my father was a software developer. Interacting with others across physical spaces through chat rooms on America Online, using digital content to complete homework assignments, and reading and contributing to early Usenet groups according to my interests (centered mostly on reading and posting about my favorite bands), everything I did online was reflective of my interests and obligations offline. I knew tacitly, then, that the online and the offline were co-constitutive.

Ethnographers of cultural practices associated with the online have defined "the field" in a number of ways. Early "cyberethnographies" focused on groups formed around a shared interest. These were often studies of online communities (e.g., Holeton, 1998; Smith & Kollock, 1999; Thomas, 2004, 2005; Turkle, 1995), Listservs, online message boards, blogs, game sites, and more that examined only the online. In the early 2000s, ethnographers recognized that the Internet is part of everyday life (Howard & Jones, 2004; Miller & Slater, 2000; Wellman & Haythornewaite, 2002) and Internet-limited ethnography fell out of vogue.

A study that focuses on online and offline contexts as two parts of one whole defines the field of ethnographic study on the basis of the research focus, in this case youth's digital composing practices. In other words, rather than depending on an a priori delimitation of the field, I explored boundaries and spaces, physical and virtual, and defined sites through participants' practices.

Because I relied on connective ethnography, the primary method of data collection was participant observation in the classroom and participants' Web spheres, collected in a field journal with artifacts including digital texts, combined with participant oral reconstruction of online travels and recursive interviews online and offline designed to create a thick description of their practices that described not just actions but social meanings. To present these kinds of social meanings, a typical ethnography might present all systems that characterize a particular group, including those associated with communication, tools to accomplish tasks, and defining behaviors (Wolcott, 1987). The problematic issue is that the Internet is, in part, a conductor for social flows, including language, tools, and behaviors. Markham (1998) explained it as a place, a tool, and a way of being. They are both "socially constructed and at the same time social conduits" (Markham & Baym, 2009, p. 11). In this same vein, Hine (2000) defined "cyberspace" as both a culture and a cultural artifact. In this view, the Internet is a communicative system itself as well as being comprised of text in its programming and at each "node" on the World Wide Web, expressed as websites. These texts, too, are often sites of interaction.

Social Groups versus Cultures and the Internet

Using the term ethnography suggests the study of not only culture but a culture, or a bounded group. Appadurai (1996) cautioned against collapsing the ideas of social groups and cultures, and called attention to the diversity within groups. He resisted the popular idea of culture as a noun, and suggested culture is "not usefully regarded as a substance, but is better regarded as a dimension of phenomena, a dimension that attends to situated and embodied difference" (Appadurai, 1996, pp. 12–13). He urged readers to consider the diversity within culture, and to use the term culture to recognize the differences among people, but not as a means of marking boundaries among people.

Thus, even within a group of people who interact frequently, the people in that group are part of many cultures, including cultures associated with religion, geographic location (and diaspora), interests and hobbies, languages, political orientation, and sexuality. When people interact frequently using the Internet, which allows for interaction across geographic and even temporal boundaries, the idea of a bounded culture is further problematic.

In this study, for example, the participants did share geography and some cultural elements. They were, at the same time, diverse in many key ways. In addition, they had exposure to many other cultures and subcultures by way of the Internet, and in those online spaces that are rich zones of connection, culture as a bounded notion is problematized. In undertaking this research, I was aware that I was studying social practices and not some static or monolithic notion of culture.

Space as Connection

Much of the research on literacy education foregrounds place and examines practices in conjunction with particular physical spaces (e.g., urban, rural, southern). Place is an organizing principle that refers to a geographical location. Space is distinct from place in that it is malleable, shifting, and is comprised of social relationships (Jimenez, 2003), a "field of relations" (Olwig & Harstrup, 1997, p. 4) rather than a location defined by physical boundaries. Theorizing around the concept of space defines it relationally as socially produced (Lefebvre, 1991). Space is not a container or "simply a static background 'upon' which human activity is played out" (Leander & McKim, 2003, p. 212). Drawing on Lefebvre (1991) and Olwig and Harstrup (1997), one may think of space as an intersection, a never-ending expanse of connections. In this view, the researcher foregrounds social relationships and social practices as well as how people navigate social spaces (Moje, 2000) instead of material place or geography (Leander & Sheehy, 2004). Such a view of space invites consideration of how and why people take up particular literacy practices, what shapes their perspectives, how they approach those composing practices within and across physical settings, and how they produce spaces in doing so.

In this study, I sought to locate three teenagers' literacy practices first in one material setting while acknowledging that social relationships, literacy practices, and discourse are constructed and exist across time, space, and other "boundaries." In making the distinction between physical places and space as connection, I did not disregard the existence or importance of material place. Material place is, after all, part of the field of relations that constitutes space (but is not wholly space itself).

The participants in this study, for example, live in a rural place with a particular history, associated cultures, and affordances and constraints. They certainly experience a material reality. At the same time, they engage in multimodal composition across socially constructed places like school, websites, clubs, and their homes. Through their multimodal composing practices, they also create spaces for themselves in which they are best friends, new friends, artists, videographers, fashion icons, or demonstrate other identity performances in differing combinations. Marginalized in the school space, students like the participants in this study, who have a history of school failure and are perceived to be "at-risk," exercise mobility and move from this deficit-space consciously or unconsciously toward spaces where they are able to demonstrate efficacy.

Jimenez (2003) described this idea of mobility across spaces:

> Social life is no longer to be seen as unfolding through space but with space, that is, spatially. Space is no longer "out there," but a condition or faculty—a capacity-of social relationships. It is what people do, not where they are. (p. 140)

The material world often influences the social world in particular ways. For example, school and district zoning creates boundaries and facilitate or constrain connections. Physical proximity and material realities such as bodies of water can constrain human mobility and thus social relationships. Underfunded schools may not have access to technologies that allow for social connections that transcend these kinds of boundaries. However, this view of space as a capacity of social relationships declares that the physical reality is not the only reality.

Technology and Space

Technology integration has been explored as a means of addressing the educational needs of rural places. In particular, researchers have approached technological tools and digital texts as a means of countering the lack of literacy resources in some rural places as well as the geographic isolation that defines some rural places (e.g., Howley, Wood, & Hough, 2011; Larson & Murray, 2008; Mitra, Dangwal, & Thadani, 2008). Access to resources and overcoming geographic limitations are issues of spatial justice (Soja, 2010).

Yet, in addition to the access to tools, resources, and other cultural capital, spatial justice also includes the subjectivities and connections available to people within and across spaces. Soja (2010) urges a critical spatial perspective to advocate for justice in contemporary society. Drawing from Lefebrve (1991), he defines space as socially produced (Lefebvre, 1991). While in this view, space is not a container or "simply a static background 'upon' which human activity is played out" (Leander & McKim, 2003, p. 212) and is thought of instead as an intersection, it still includes and accounts for the material and physical aspects of place. In this view, what is foregrounded are social relationships and social practices as well as how people navigate social spaces

(Moje, 2000) but the material world is not excluded. The analysis presented in this chapter first locates the participants' literacy practices in one material setting while acknowledging the ways that social relationships, literacy practices, and discourse are constructed and exist across time, space, and other "boundaries." This involves a mapping of the contours, flows, and nodes within relevant spaces including texts, social connections, material objects and physical places.

Materiality

Perhaps as a reaction to a twenty-year focus in the field on the semiotic, which deals with immaterial abstraction, literacy research has most recently taken a turn toward the material. This turn involves addressing ontological questions about the nature of literacy practice that includes a focus on the physical artifacts people engage in these endeavors. Because objects can be viewed as the "stuff" that people use to make meaning (for example, a particular brand of shoe can be used to signal belonging to a particular culture), this focus on so-called "artifactual" literacy (Pahl & Rowsell, 2013) can be seen as in line with multimodality. As Pahl and Rowsell (2013) clarify, "artifacts bring in everyday life. They are material, and they represent culture. They can link literacy, multimodality, and material culture together by focusing the lens on meaning-making that is situated and material" (p. 268). In this view, objects and other artifacts become the representational modes that people use in their contexts of being to make meaning.

Cultural and material studies such as those of Hurdley (2006) provide examples of analyses of tools for producing texts but also the materiality of texts. Hurdley's (2006) ethnographic study of mantelpieces in homes showed the ways which these material artifacts highlight particular aspects of social practice, such as how objects can demonstrate or perform certain aspects of our selves. A statue on a mantelpiece, for example, might represent a daughter's love for the mother to whom she gave it, and it might also represent that mother's artistic taste or style.

Much has been said about the materiality of our tools in relation to literacy practice, but it might seem counterintuitive to analyze the materiality of digital texts. However, they have not only particular physical qualities such as those that give them structure, but they also have assets that make them a discrete product, per se. They are further physical in the sense that they can travel across spaces like those of home and school. They are observable and measurable; they have physical properties. A focus on the material qualities involved in mobile phone-based literacy practice (such as the tool of the mobile phone or objects represented within texts produced by the phone) can help researchers to understand these literacy practices in a more situated way, since tools appear and are taken up at a particular point in time and space and within a particular cultural milieu.

An emphasis on the material world is present in mobile phone-based photography. Physical aspects are present in the regard for angle, color, lighting and other physical properties that are considered in youth digital composing with mobile phones. Youth photograph and post their homes, classrooms, friends, family members, and possessions. In mobile phone photography, in particular, there is a real regard for the physical properties of things and the material culture of home, school, and other spaces of youth's lived experience is observable. In this way, youth enact transmediation (Siegel, 1984), turning physical objects into digital artifacts or digital texts through mobile phone photography. This practice provides a means of sharing their physical worlds with their social connections. The physical and material aspects of the spaces they inhabit are often considerations within digital composing practice.

Embodiment

Recent research considering the spatiality of literacy practice has focused on the movement of bodies through and across spaces, and this research often considers the way literacy practices are embodied (e.g., Johnson & Vasudevan, 2012; Leander & Boldt, 2013; Wohlwend, 2008). This can include the sensory qualities that we feel as users (Ingold, 2007; Pink, 2009) in reaction to or in conjunction with the physical world. Digital texts have qualities that are perceived by the senses or even suggest, through synesthesia or otherwise, perceptually processed information. It can also include the ways that bodies show up within literacy practice. Youth often use their bodies as a "local, daily text" (Johnson & Vasudevan, 2012, p. 36). They use clothing, hairstyles, accessories, tattoos, piercings, and other adornments to dialogically interact with a world of others who "read" and make meaning of their bodies. They can use their literacy practices as a means of representing their bodies across spaces (consider, for example, the selfie shared on social media).

Materiality and embodiment collide, too, when we think about the phone as a tool for giving materiality to emotions. Emotions are immaterial but when we use the phone to mediate them, they are given physicality and an ability to travel and be heard and seen and otherwise perceived or felt by others. They are given physical form and sent out into the world. This materiality could occur in other different combinations as well. For example, perceiving something in the world could cause an emotional reaction that is felt and that might spur one to post a text. Such meaning is made in dialogue with the viewer or audience. Pink (2007) claims that: "visual images are made meaningful by the subjective gaze of the viewer, and that each individual produces these photographic meanings by relating the image to his or her existing personal experience, knowledge and wider cultural discourses" (p. 82). It stands to reason that digital postings about emotions are a means of connecting with others.

Early ethnographic studies of the Internet explored this concept of embodiment as well. For example, Thomas (2004) created virtual reality space Palace to conduct an ethnographic study of online communities in work with a virtual world. In this virtual world, youth communicated through virtual embodied selves or avatars. Because Thomas conducted research including interviews and participant observation as an avatar, the researcher was able to experience the context of the research in a rich and indeed more embodied way. Thomas not only studied how participants created their digital selves including digital bodies through the semiotic resources available in the virtual world as an observer, but in writing digitally in this way, Thomas lived out the practice.

In the early days of the Internet, it was predicted that because communication was disembodied, race, gender, and other differences would be irrelevant or disappear. Instead, people use the Internet to make sense of their racial and gendered identities and form community based on these identities online (Byrne, 2008a, 2008b; Everett, 2004; Nakamura & Chow-White, 2011). The idealized, race-free version of the Web sprung from the early text-based Web; with the multimodal Web, Nakamura (2008) maintained that one represents and plays with race and other markers of identity more than previously.

The next section uses the data from the study of the participants' mobile-phone based composing practices to illustrate some of the concepts discussed in the prior section. Through a case-based organizational method, the entangled concepts of space, place, materiality and embodied practice are evident.

That is to say, these concepts are difficult to represent individually since they are interrelated and interdependent.

Mobile, Embodied Composing across Spaces

All three of the participants in this study harnessed an array of symbols and semiotic affordances to communicate in and across digital and physical spaces. Access to the Internet and, in turn, audiences, readily available also expanded their communicative potential. They each used their mobile phones and their attendant affordances to accomplish their individual moment-to-moment goals.

Eli

Eli had just learned to drive, but he did not have a car of his own (he was working his new job to save up to buy one) and over the course of the study, I often saw him walking around different parts of Brookeville or lingering in front of the school near the road, always talking on his phone. He did not have Internet access at home outside of his phone, although he did have an older-model computer at home to which he had access for typing school assignments. He was allowed to take the iPad from White Birch home at night, but did not use it at home because it had reduced functionality imposed by the school, which sought to regulate the use of the iPads, and because there was no Internet access at home.

I asked Eli early on when and how he accessed Twitter:

Eli: Yeah, usually on my phone.
Julie: You guys can't get on the Internet on the phone at school, can you?
Eli: We ain't got no service here.
Julie: I know! Right when I pull in the parking lot the cell phone goes dead.
Eli: I don't know what they did, but it works. *Laughs*
Julie: *Laughs* Um, so you can't really do anything on Twitter or anything like that?
Eli: At school, no.
Julie: And then they took, they made it to where you can't get on [Twitter] the iPads, right?
Eli: No, they did something to that, too. We did in the beginning of the year, but they, I guess they switch it up or something.
Julie: Yeah. That makes sense. You usually get on it at home?

> Eli: Yeah, but like um when I'm waiting for someone … because I'm not at home most of the time, but when I am, I'm usually on Twitter, like, when I'm doin' somethin' and I'm bored, I'll go on Twitter.

Eli engaged with digital media across spaces and moved through "multiple, simultaneous space-time contexts … coordinated and produced through activity" (Leander & McKim, 2003, p. 238). His practices moved fluidly across spaces. He often posted pictures to his social media spaces he had taken while at school. He engaged in school-like practices outside of school such as composing narratives through photography. In school, he explored the possibilities for life and he did the same through his digital literacy practices (Moje, 2000). In and out of school, he used his phone to call upon the Google search engine as a means of locating the information he needed to accomplish moment-to-moment goals (although, as I observed, he had to move near a window to get a cell phone signal in the classroom, much to his teacher's chagrin). However, he didn't move into global cosmopolitan spaces. In fact, he preferred to interact only with people he knew from the physical world in online spaces. Early on, he described how strange he found the practice of connecting with strangers in the online space after someone he did not know communicated with him on Twitter:

> Eli: I text somebody I didn't even know, I'm like, what is you talkin' about, cause I don't even be hanging with you. He was like, that's Twitter man.

Technology affords the mobility needed to move into the global space, but as Frith (2012) writes, "how much control individuals have over their experiences while mobile is a function of how legible particular spaces are for the people who move through them" (p. 134). Eli was unprepared to communicate in the "global ethnoscape" (Appadurai, 1996) because the Internet he used was scaled down. In his own digital literacy practice, he scaled it by choosing his social connections and thus creating the space. At school, the Internet was scaled via firewalls intended to protect students. He was also unprepared because he needed instruction about the nature of digital writing contexts and about geographic and temporal scalability as well as the persistence of digital texts in a context of transnational global flows if he was going to participate in that space (Castells, 1996). In school, however, this was not part of the curriculum. Technology was mainly used for word processing and assignment submission via an application called Edmodo (course management software).

Eli's embodied presence in online spaces showed not only the ways that he represented his physical being and markers of identity in the online space, but also how he used digital media to create a space. Eli often posted photographs that expressed affiliation with a neighborhood gang. Because he was often in trouble in the schooled space, it was not surprising to me that he acted out a "bad boy" narrative online. The way he was positioned at school was reflected in his digital literacy practice. Moje (2000) asserted that teens shut out from the dominant "story" because of ethnicity, race, or social class used gang-related literacy practices to write "themselves into the world" (p. 652). Literacy practices involving gang practices "may serve as tools for transforming thought and experience in the lives of marginalized youth" (Moje, 2000, p. 652). I recognized Eli's digital gang affiliation practices were a form of agency, of exploring an identity and of making himself known. Eli wanted to do well in school and strove to stay current in his work in Ms. Harrison's class even when he was suspended or otherwise had to miss school. By his account, he was a good student and had to be a good student so he could continue to wrestle at Whale Branch. He described wanting to make his coach and parents proud of him, and excelling in school was a way to do that. He believed school was important, and considered being educated to be part of what defined success. He often posted photos of himself dressed impeccably for school, wearing one of the many ties from his collection. One such photo carried the caption: "@edumacated swag, son." Through clothing and other material vestiges, Eli used his body and physical materials to create counterspaces and to represent his identity.

On Twitter, one can post tweets to all of their followers or direct tweets to particular audiences using "@." Eli rarely posted "@" messages on Twitter. Most of his tweets were entirely open. He would post @ messages if he was responding to content a friend had already posted. Most of his tweets were not of this sort; instead of being directed toward one particular user, they were open on his page. These open tweets had an affective quality, in that many of them related to moods, feelings, beliefs, and attitudes. For example, the last seven original tweets Eli posted that were not directed toward any one user were all affective:

> I hate 6ein so6er
> Thiz iz how fukin 6oared i am! (attached picture of a pyramid made of empty soda cans)
> Ya im definently not goin 2 skool 2day
> Ma lil nigga growin up so damn fasssssstttt (attached a picture of his baby niece)

> Suspended from skool again,i needa learn 2 controle ma self
> Damn man I juz kant konvince ma self I dont lov her
> Man ma sista n her hus6and n ma kuzin n her 6a6y daddy in here talkin 6out dare relationshipz ... Damn i needa a girl

Dutro (2008, 2009) wrote about the exclusion of the affective from schooled literacy spaces, pointing to the way traditional curriculum and teaching leave no room for the emotional side of the self in academic spaces. Perhaps Eli used Twitter to express his feelings because he did not have the space at school. The availability of the phone facilitated by its small size and light weight as well as its centrality to his communicative practice (he had to have it to communicate with his mother, who he depended on to drive him to his many activities throughout the day) allows for instantaneous release of emotions via photographs and linguistic text.

In one focal group interview, the participants in this study described a practice related to emotional release common on Twitter they called "subtweeting." The term is short for "subliminal tweet." By their account, the subliminal tweet is a means of talking about someone without actually calling them by name or tagging them in the post. The practice was a passive-aggressive one that allowed those who engage in it to talk behind someone's back openly. Eli did this quite often. For example:

> 6itch ur life really trully kant 6e dat damn if every pik i seen of u u smilein in
> Some 6ody needa tell deez girlz day are not cuuuuuutttttte
> Damn nigga u tweet a lot shyt get a life nigga

When Eli explained that these tweets were merely a means of venting and the choice to withhold the names of those to whom he is referring was so as to avoid "drama." "You be feelin better after you write that," he explained to me.

John

Like Eli, John used his phone as his primary means of access to the Internet. He told me, "I get on a lot in school, in church, and in the car, or at home." John was much more active online than the other two participants in the study, and had accounts from which he posted content on Twitter, Instagram, and YouTube. John marked his movement across web spaces by embedding hyperlinks to his various profiles on the sites where he interacted. Halfway through the study, I noticed that he and other youth had added the letters "IG" and a handle or username to their Twitter profiles.

Julie: What is "IG"? Would somebody …
John: Instagram.
Julie: Oh, well why would … ok so that peo … so that you could kinda link your Instagram to …
John: Nah, when you put it …
Julie: To your Twitter?
John: When you put that on Twitter it's telling people that that that's your IG name so go follow it.

To address this need for unity across his sites of interaction, John, along with other youth in his Web sphere, created his own conventions (using an abbreviation of Instagram, IG, attached to their Twitter names). This allowed them to manage the vast space of the Web by creating their own infrastructure to make themselves locatable online. In this way, he scaled the Web to a size manageable for him and marked his movement across spaces.

Like Eli, the majority of John's original (i.e., not retweeted) tweets were related to emotions. Like Eli, his mobile phone represented a tool for giving shape to and communicating his inner workings. It was a tool he could use to mediate his emotional states as well as display them. Both participants used the mobile phone as a means of managing emotions and were learning about how their emotions were similar to or different from others' as they interacted in the digital spaces where similar emotional mediation was happening. The phone also thus facilitated social connection as they shared their own emotions with others in their web spheres.

Like Eli, too, John engaged in embodied digital composing practice. John had a penchant for fashion, and often mentioned his "swag" or "swagger" in his online postings. Instagram, a visual medium, offered him a means of displaying his fashion sense using his body and various articles of clothing. He often used a smartphone app called Pic Stitch to meld several photos together into one picture, giving what I coded his "fashion" photos a sense of unity. In one such artifact, he labeled a series of photos "Turn up Thursday" and posed with a Styrofoam cup to his mouth. "Turn up" is a popular culture reference to partying that is often signified with a Styrofoam cup. "Turn Up" is also the title of a song by one of the most beloved rappers of Ms. Harrison's ninth-grade English class, 2Chainz. (During my observations, I often heard mention of 2Chainz, and even Ms. Harrison joked about the students' love of 2Chainz).

John used the physical object, this Styrofoam cup, to show his affiliation with or knowledge of hip hop subculture. He often used physical objects to communicate such messages.

John also used his phone to record videos, which he then posted online to YouTube. He maintained his own YouTube channel where he posted seven original hip-hop songs, a substantial body of work, considering he had to come up with the music and beats as well as write the raps, and then record, edit, and upload the songs. In these videos (and in some cases uploaded audio files matched with a slideshow of still images), he uses clothing to demonstrate affiliation with hip hop culture and to make a space for himself as an emerging musician. In one video, for example, he includes an introductory frame featuring himself and his friend with whom he wrote and recorded the song. They chose pseudonyms and seemed to play characters in the rap verses, which feature the typical themes and tropes of the rap-music genre. The primary theme on which they fixate is their fashion sense, singing, "We swagged up, don't act like you ain' notice/She say she like my style, so I keep it goin/These girls goin wild, so I keep it goin." A vocal track loops in the background of John's friend yelling out, "Steez!" (a slang term for style). The performance of style, which relies on artifacts of the material world, is central to the creation of this space as artist. As the song and video progresses, different photographs appear on the screen of John and his friend in various fashionable outfits, standing in poses commonly featured in rapper photo shoots, on album covers, and in music videos. They use their bodies in this way to embody this identity of rapper or hip hop artist. John used a free program he downloaded from the Internet to create the video. He explained he was able to drag and drop the photos they had taken with their phones and emailed to themselves to transfer them to the computer to create the video.

These videos show the ways that John created and remade spaces using embodied literacy practice. For example, the videos feature an array of jewelry and accessories worn in ways that depart from the more staid dress I had observed at White Birch High School. He is thus able to use materiality to represent a new identity and open up a space for himself to enact a different positionality than that which he experienced in the schooled literacy space (that of student in a failing school with failing scores on federally-mandated tests).

Aaliyah

Aaliyah's online space of choice was Twitter. She told me she uses Twitter because it is a separate space from her parents' online space. "Facebook is everybody," she explained. She went on to cite that even her grandfather has a Facebook: "They're parents and so they've heard about Facebook from other people and they know what it is." She seemed to want a separate space from the adults in her life.

Aaliyah told me she first set up the Twitter account because she moves around a great deal. White Birch was the third school she had attended that year, and she anticipated another move soon, telling me her mother had been talking about wanting to move the family to a larger city nearby. Twitter was her way of staying in touch with friends whom she was unable to see as a result of moving around. Aaliyah recognized and harnessed her ability to produce a new space with digital tools (Lefebvre, 1991). When she was at home, she could hang out with her family, but also be "with" her friends because "multiple, simultaneous space-time contexts are coordinated and produced through activity" (Leander & McKim, 2003, p. 238).

Aaliyah comported her self differently in the different spaces she inhabited. In all of my classroom observations, Aaliyah was Ms. Harrison's helper, the kind of student every teacher appreciated. She sat near her teacher's desk, and whenever she noticed that papers needed to be distributed or materials needed to be collected, she rose automatically to assist. Although many of the students in the class were boisterous and vocal, Aaliyah was quiet. She was constantly watching and listening, and seemed to aim to please her teacher. Ms. Harrison informed me that she was good friends with Aaliyah's mother and that they attended the same church and talked there often about Aaliyah's performance in her classroom. This may have been why Aaliyah was so invested in her English class. In my talks with Aaliyah, she insisted that she was a "class clown" and that she was always goofing off with her friends. This was a side of her I never saw in Ms. Harrison's class. There, she was always focused. Aaliyah said that everyone expects her to be a certain way because of her class-clown persona, but she is actually smart and does well in school. She had been on the honor roll throughout middle school, just before starting at White Birch High School. Aaliyah's dedication to education might have reflected that she had watched her mother struggle to get an education. Her mother was 17 years old when she had Aaliyah, and just the year before had got her high school diploma through a local human-development center.

"Every morning you wake up and look at your daughter, you're more determined," her mother said of what sustained her as she struggled to earn her high school credentials. I talked to Ms. Harrison about the disparity in Aaliyah's perception of herself and what I saw; Ms. Harrison agreed that Aaliyah was an ideal student in her classroom, hardworking and cooperative. However, Ms. Harrison admitted that she felt young people acted differently depending on who was nearby.

Aaliyah seemed to have a sense of the embodied nature of her digital composing practice. That is to say, she seemed to understand the interactive nature of online literacy practice as central to the space in a way that made her feel her own embeddedness in the online space. Though she loved books, she made the distinction that the text online was alive: "With books, I'm imagining it, but with technology, I'm in it, it's more fun." She described the way she could connect with people all over the world on Twitter over mundane daily happenings. "I might be like, 'In class and this lady is talking a lot!' or, 'I think I just failed this test' and I'll get like a bunch of followers off that." She recalled posting about some food she was eating, and how the tweet initiated a conversation about food with someone in London. She knew this was a space of global flows (Appadurai, 1996).

Aaliyah's literacy practices were comprised of a mix of print text and digital text, as well as embodied and visual literacies. She told me she reads many books, like *Twilight*, *The Hunger Games*, and *Harry Potter*. "I love books but I've grown up with technology. So it's not like I'm gonna just read a book after a book. That's just too much," she explained. "With books, I'm imagining it, but with technology, I'm in it, it's more fun." She experienced her online composing practice as an embodied practice that put her squarely in the middle of the action, rather than outside of it.

Aaliyah had an understanding of the ways in which she was afforded freedoms or constrained in the different spaces she inhabited. Online, she was able to communicate more freely than she could at school. She explained that she liked the fact that she had some freedom from the strict conventions of schooled English when she wrote in the online space. She felt some of the grammatical rules were absurd and baseless. They also ran counter to her own linguistic heritage, one that was rooted in a creole language with a morphology and syntax distinct from that of the Standard American English taught in school. In this way, she was less free to enact her identity in the schooled space.

Julie: What about the spelling [online]? How it's not like "correct English?"
Aaliyah: Ugh, I hate correct English!
Julie: Why?
Aaliyah: Because when I get home, I use ain't so many times, I use a lot of slang … and I think that's another thing with Twitter how it's different from school is you always have to use proper grammar, and I hear that every time I'm in school and I hate it so bad.
Julie: Oh, because on your papers there will be marks that say …
Aaliyah: Yes! It will be like, the "sp" like "spelling" and I'm like, oh, why couldn't you just let me have that! Like, I'll put the "to" with the 2 o's, and I'll put the one o, and I'll be like, that doesn't even make sense! Please explain to me, you're the teacher, explain to me, because that is …
Julie: Yeah, it's faster to write t-o.
Aaliyah: Or just write the number "2."
Julie: Yeah.
Aaliyah: Or just put it how you pronounce it. Instead of saying "listening" like instead of "listeninG" (emphasizes the "G" sound) I put "listenin'" and I'll spell it "l-i-s-n-i-n" instead of putting the whole word.
Julie: So you just like it because it's faster or easier, or why would you rather write in slang or shorthand?
Aaliyah: Because, those are the words that come to my mind. It's like when your teacher tells you not to write everything that comes out of your mind because it doesn't come on the paper right, and I do that A LOT, and so, especially when I'm thinking about something and I have a lot to say about it, I'll put it, I'll be writing so fast that I cut off certain things and get into a whole different subject.

She seemed to appreciate the stream-of-consciousness approach to communication that would allow for the unmediated expression of her thoughts and mental meanderings. Writing online is best unplanned, unmitigated, unfettered by attention to grammatical rules for which she did not understand the logic, like homonyms. She also seemed to enjoy that she could more fully enact her identity in the online space.

Aaliyah understood her power to use media to socially produce a space (cf. Lefebvre, 1991) where she could express her quirky side and "act stupid" as she put it.

Aaliyah: It's a messaging thing. So, if I wanted to talk to somebody about something stupid, why would I want to do that in class, I'm supposed to be, sounding kind of smart here, but, you know, you're supposed to be getting your education and the teachers wouldn't allow that. It's more like a passing notes thing.

Julie: Ok, yeah. The purpose is different.
Aaliyah: The whole thing is different!

Her conceptualization of the nature of space is that it can be created, traveled to, and inhabited in a way that reflects one's own agency.

Like Eli and John, Aaliyah also used digital media as a means of connecting with others through sharing of emotions. Aaliyah used the practice of favoriting a post on Twitter as a way of communicating agreement, alignment, or connection with another person through the touch of a button. To favorite a post, she would tap the star icon under the post. She explained it thusly:

Aaliyah: A lot of people, like a lot of tweets that you put on Twitter, a lot of people are feeling at the same time. So, like say, somebody across the world could be like, "Man, I'm washing my hands," or something. Or, "I'm using the bathroom." And you could be saying the exact same thing they just did, and tweeting it, and then they look at it and think, "You know what? I'm going to put that as my favorite because I feel that way right now."
Julie: Oh, I didn't know that. What kind of stuff would you favorite on here?
Aaliyah: A lot of stuff, like, basically what I said, how like, if I'm feeling some type of way and I didn't think about putting it on Twitter, and I was just thinking about it, somebody could put it on there and I would favorite it.
Julie: So if I went and looked at your favorites, I would just see like different … What would I see?
Aaliyah: People would say things like, let me think of one I recently favorited … umm, there was this one about this girl named Danielle, and she's my best friend. We've been friends since first grade. And she said something about, um, love, and it was like, "I love you and I gave my all to you." And then I favorited it, because, you know, you have your little crushes on the boys and everything, so I favorite that one.
Julie: Ok, so just something that you related to, so then you favorite it.
Aaliyah: Yeah, that's exactly it. Something that you relate to.

She could communicate her agreement, support, or understanding through the press of the favorite key and connect with others through doing so.

Similarly, Aaliyah told me she was often inspired or delighted by content she consumed online. She saw it as a way to regulate her emotional state. For young people like Aaliyah, digital media offer a means of asserting control over the fluctuating emotional states of adolescence. She followed an account, for example, called "Keep Calm," which she explained to me was "a person that, they put a lot of, like 'relax, it's not going to be that big' stuff." She liked to read these posts, ostensibly, to help her relax and put the events of her life into perspective. She liked Twitter, because, as she put it,

> It's for the people that always have something good to talk about. Maybe the people that have all the followers and everything ... a lot of people that put quotes up, the quote of the day or tip of the day. People like that or just interesting people. It's just stuff that you'll laugh at that you wanna hear. Something you would wake up in the morning and something like, to keep you happy.

She might read her Twitter feed in the morning to be exposed uplifting posts from "people that always have something good to talk about." She might read it to become privy to posts that will make her laugh and keep her happy.

Aaliyah used the Internet to keep track of her friends, who were distributed across the region. Although geographic boundaries prevented her from meeting and conversing with her friends in different physical spaces, she created a digital space where she could converse with friends in spite of the physical distance. She agentically created a space for herself and her friends, apart from her parents and family members, where she could be "stupid" and "post stupid stuff." It was her own space where she was in power and where she curated the digital museum of herself. When she was at home, she could hang out with her family, but also be "with" her friends because "multiple, simultaneous space-time contexts are coordinated and produced through activity" (Leander & McKim, 2003, p. 238).

Aaliyah created space by harnessing the serendipitous relationships in social spaces extending beyond her immediate material location (e.g., her dispersed friends, as well as Twitter users as far away as London). She was able to learn socially through discussion and to learn from social interaction without social proximity. She produced her own social spaces (Lefebvre, 1991). She learned about communicative norms in the online space as well as snippets of cultural information, as she demonstrated through her interaction with a Londoner, through the production of these social spaces.

Aaliyah's expressiveness varied in different spaces. At school, she was quiet and reserved. Online, she was boisterous and silly. Her handle included derivatives of the words "psychotic" and "crazy." She saw the online as a safe space where she was among friends, a space she created through choosing the social connections that comprised it (Jimenez, 2003). She saw her space on Twitter as her own space where she displayed and posted content that reflected her personality, beliefs, sense of humor, and worldview. She adapted to different discursive spaces as needed. She relied on writing to communicate in these different spaces on and offline, but approached communication differently depending on "activities, identities, and power at a particular moment in time" (Siegel & Panofsky, 2009, p. 101).

Discussion

Even when creating new spaces for enacting identity, connecting with others, engaging in emotional release, or just hanging out, youth composing practices are not fragmented but instead are continuous across spaces. Mobile phone-based composing practices are rooted in their offline social worlds. For example, the youth in this study moved away from marginalizing, deficit spaces into spaces where they could more fully participate. In addition, they shared their physical spaces and material worlds by making them digital and shareable through mobile phone-based photography. Their composing practices, drawing on affordances within both the physical and digital spaces, were thus continuous and fluid.

Phones represent near-constant companions for youth. Further, youth interact in intimate ways with these technologies, pressing them to their faces, sleeping with them, carrying them close to their bodies in pockets and handbags. This continuous proximity to the body makes the phone, in a way, another extremity that changes the way that youth can interact with and experience the world and experience their own inner landscapes as well.

In addition to being intimately related to the material phone, youth use their mobile phones also to foster a sense of intimacy with others and a connection to their social worlds. For example, the youth in this study used their phones to share affective communication with different audiences including close friends, enemies, strangers, and/or more amalgamated and less defined groups of followers. They made their inner states known to others in this way. These new kinds of affective communication will and do shape offline social interaction in salient ways.

As youth continue to create and inhabit new spaces, new methodological tools that allow researchers to follow or trace the mobilities of youth would be beneficial. For example, the participants in this study leveraged the temporal properties of technologies (e.g., the ability to "capture" a moment through photography, the affordance of the technology to provide a means of display that allows for revisiting—a sort of digital memory). New ways to represent data that can represent time and space relationships are needed. What is possible now is only a fragmented connective ethnography, but with digital tools that make visible youth travels though time and space, researchers can learn more about how youth take up and make meaningful digital tools and their travels through these created spaces.

Viewing mobile phone-based composing practices through a spatial lens also offers an opportunity to recognize the social uses of mobile phone-based literacies. Instead of a focus on texts and composing as solely a communicative process, we can begin to see the social, emotional, and indeed even embodied uses of mobile composing practices for youth as they interact with the physical and immaterial worlds they inhabit. We begin to see literacy practices as more than cognitive accomplishments, one-time performances that are representative of skills and knowledge in a particular space and time, and can recognize the mobility, fluidity, and spatial complexity of these practices.

References

Appadurai, A. (1991). Global ethnoscapes: Notes and queries for a transnational anthropology. In R. G. Fox (Ed.), *Recapturing anthropology: Working in the present* (pp. 191–210). Santa Fe, NM: School of American Research Press.

Appadurai, A. (1996). *Modernity at large: Cultural dimensions of globalization*. Minneapolis, MN: University of Minnesota Press.

Baron, N. (2008). *Always on: Language in an online and mobile world*. New York, NY: Oxford University Press.

Blinka, L., & Smahel, D. (2009). Fourteen is fourteen and a girl is a girl: Validating the identity of adolescent bloggers. *CyberPsychology & Behavior, 12*, 735–739. doi:10.1089/cpb.2009.0044

boyd, d. (2007). Why youth (heart) social network sites: The role of networked publics in teenage social life. In D. Buckingham (Ed.), *Youth, identity, and digital media* (pp. 119–142). Cambridge, MA: MIT Press.

boyd, d. (2008). *Taken out of context: American teen sociality in networked publics* (Unpublished doctoral dissertation). University of California-Berkeley, School of Information, Berkeley, CA.

Brouwer, L. (2006). Giving voice to Dutch Moroccan girls on the Internet. *Global Media Journal*, 5(9). Retrieved from http://dare.ubvu.vu.nl/bitstream/handle/1871/25786/194682.pdf

Buckingham, D. (Ed.). (2008). *Youth, identity, and digital media*. Cambridge, MA: MIT Press.

Byrne, D. N. (2008a). The future of (the) "race": Identity, discourse, and the rise of computer-mediated public spheres. In Everett, A. (Ed.), *Learning race and ethnicity* (pp. 15–38). Cambridge, MA: The MIT Press.

Byrne, D. N. (2008b). Public discourse, community concerns, and civic engagement: Exploring Black social networking traditions on BlackPlanet.com. *Journal of Computer-Mediated Communication, 13*, 319–340.

Castells, M. (1996). *The rise of the network society: The information age: Economy, society, and culture* (Vol. I). Oxford: Blackwell Publishers.

Clifford, J. (1992). Travelling cultures. In L. Grossberg, C. Nelson, & P. Treichler (Eds.), *Cultural studies* (pp. 96–116). New York, NY: Routledge.

Davies, J. (2011). Discourse and computer-mediated communication. In B. Paltridge & K. Hyland (Eds.), *Companion to discourse analysis* (pp. 228–243). London: Palgrave.

Davies, J., & Merchant, G. (2009). *Web 2.0 for schools: Learning and social participation*. New York, NY: Peter Lang.

Dutro, E. (2008). "That's why I was crying on this book": Trauma as testimony in children's responses to literature. *Changing English, 15*, 423–434. doi:10.1080/13586840802493076

Dutro, E. (2009). Children's testimony and the necessity of critical witness in urban classrooms. *Theory into Practice, 48*, 231–238. doi:10.1080/00405840902997519

Everett, A. (2004). On cyberfeminism and cyberwomanism: High-tech mediations of feminism's discontents. *Signs, 30*(1), 1278–1286.

Friedman, T. (2005). *The world is flat: A brief history of the twenty-first century*. New York, NY: Farrar, Straus and Giroux.

Frith, J. (2012). Splintered space: Hybrid spaces and differential mobility. *Mobilities, 7*, 131–149.

Hine, C. (2000). *Virtual ethnography*. London: Sage.

Holeton, R. (1998). *Composing cyberspace: Identity, community, and knowledge in the electronic age*. Boston, MA: McGraw-Hill.

Howard, P. N., & Jones, S. (Eds.). (2004). *Society online: The Internet in context*. Thousand Oaks, CA: Sage.

Howley, A., Wood, L., & Hough, B. (2011). Rural elementary school teachers' technology integration. *Journal of Research in Rural Education, 26*(9), 1–13.

Huffaker, D. A., & Calvert, S. L. (2005). Gender, identity, and language use in teenage blogs. *Journal of Computer-Mediated Communication, 10*. doi:10.1111/j.1083–6101.2005.tb00238.x

Hull, G., & Schultz, K. (2001). Literacy and learning out of school: A review of theory and research. *Review of Educational Research, 71*, 575–611.

Hull, G., & Schultz, K. (Eds.). (2002). *School's out! Bridging out-of-school literacies with classroom practice*. New York, NY: Teachers College Press.

Hurdley, R. (2006). *Dismantling mantelpieces: Consumption as spectacle and shaper of self in the home* (Unpublished PhD thesis). Cardiff: Cardiff University.

Ingold, T. (2007). Materials against materiality. *Archaeological Dialogues, 14*(1), 1–16.

Ito, M., Okabe, D., & Matsuda, M. (Eds.). (2005). *Personal, portable, pedestrian: Mobile phones in Japanese life*. Cambridge, MA: MIT Press.

Jimenez, A. (2003). On space as a capacity. *Journal of the Royal Anthropological Institute, 9*, 137–153. doi:10.1111/1467–9655.t01-1-00008

Johnson, E., & Vasudevan, L. (2012). Seeing and hearing students' lived and embodied critical literacy practices. *Theory into Practice, 51*, 34–41. doi:10.1080/00405841.2012.636333

Jones, R. H. (2001, November). *Beyond the screen: A participatory study of computer mediated communication among Hong Kong youth.* Paper presented at the annual meeting of the American Anthropological Association, Washington, DC.

Jones, R. H. (2005). Sites of engagement as sites of attention: Time, space and culture in electronic discourse. In S. Norris & R. H. Jones (Eds.), *Discourse in action: Introducing mediated discourse analysis* (pp. 141–154). London: Routledge.

Lam, W. S. E. (2000). Second language literacy and the design of the self: A case study of a teenager writing on the Internet. *TESOL Quarterly*, *34*, 457–482. doi:10.2307/3587739

Lam, W. S. E. (2004). Second language socialization in a bilingual chat room: Global and local considerations. *Language, Learning and Technology*, 8(3), 44–65.

Lam, W. S. E. (2006). Culture and learning in the context of globalization: Research directions. *Review of Research in Education*, *30*, 213–237. doi:10.3102/0091732X030001213

Lam, W. S. E. (2009). Multiliteracies on instant messaging in negotiating local, translocal, and transnational affiliations: A case of an adolescent immigrant. *Reading Research Quarterly*, *44*, 377–397. doi:10.1598/RRQ.44.4.5

Larson, R. C., & Murray, E. (2008). Open educational resources for blended learning in high schools: Overcoming impediments in developing countries. *Journal of Asynchronous Learning Networks*, *12*(1), 85–103.

Leander, K., & Boldt, G. (2013). Rereading "a pedagogy of multiliteracies": Bodies, texts, and emergence. *Journal of Literacy Research*, *45*, 22–46. doi:10.1177/1086296X12468587

Leander, K., & Lovvorn, J. F. (2006). Literacy networks: Following the circulation of texts, bodies, and objects in the schooling and online gaming of one youth. *Cognition and Instruction*, *24*, 291–340.

Leander, K., & McKim, K. K. (2003). Tracing the everyday "sitings" of adolescents on the Internet: A strategic adaptation of ethnography across online and offline spaces. *Education, Communication & Information*, *3*, 211–240. doi:10.1080/14636310303140

Leander, K., & Sheehy, M. (Eds.). (2004). *Spatializing literacy research and practice.* New York, NY: Peter Lang.

Lefebvre, H. (1991). *The social production of space.* Cambridge, MA: Blackwell.

Lemke, J. (2006). Toward critical multimedia literacy: Technology, research and politics. In M. C. McKenna, L. D. Labbo, R. D. Kieffer, & D. Reinking (Eds.), *International handbook of literacy and technology* (Vol. 2, pp. 3–14). Mahwah, NJ: Lawrence Erlbaum.

Ling, R. (2004). *The mobile connection: The cell phone's impact on society.* San Francisco, CA: Morgan Kaufman.

Livingstone, S. (2008). Taking risky opportunities in youthful content creation: Teenagers' use of social networking sites for intimacy, privacy and self-expression. *New Media & Society*, *10*, 393–411.

Manovich, L. (2001). *The language of new media.* Cambridge, MA: MIT Press.

Markham, A. (1998). *Life online: Researching real experience in virtual space.* Walnut Creek, CA: Rowman Altamira Press.

Markham, A., & Baym, N. (2009). *Internet inquiry: Conversations about method.* Thousand Oaks, CA: Sage.

Mazzarella, S. (2005). *Girl Wide Web: Girls, the Internet, and the negotiation of identity*. New York, NY: Peter Lang.

Miller, D., & Slater, D. (2000). *The Internet: An ethnographic approach*. Oxford: Berg.

Mills, K. A., & Comber, B. (2013). Space, place and power: The spatial turn in literacy research. In K. Hall, T. Cremin, B. Comber, & L. Moll (Eds.), *International handbook of research in children's literacy, learning and culture* (pp. 412–423). London: Wiley-Blackwell.

Mitra, S., Dangwal, R., & Thadani, L. (2008). Effects of remoteness on the quality of education: A case study from north Indian schools. *Australian Journal of Educational Technology, 24*(2), 168–180.

Moje, E. (2000). To be part of the story: The literacy practices of gangsta adolescents. *Teachers College Record, 102*, 652–690.

Moje, E. (2004). Powerful spaces: Tracing the out-of-school literacy spaces of Latino/a youth. In K. Leander & M. Sheehy (Eds.), *Spatializing literacy research and practice*. New York, NY: Peter Lang.

Nakamura, L. (2008). *Digitizing race: Visual cultures of the Internet*. Minneapolis, MN: University of Minnesota Press.

Nakamura, L., & Chow-White, P. (2011). *Race after the Internet*. New York, NY: Routledge.

O'Brien, D. G. (1998). Multiple literacies in a high school program for "at-risk" adolescents. In D. E. Alvermann, K. A. Hinchman, D. W. Moore, S. F. Phelps, & D. R. Waff (Eds.), *Reconceptualizing the literacies in adolescents' lives* (pp. 27–49). Mahwah, NJ: Erlbaum.

O'Brien, D. G. (2003). Juxtaposing traditional and intermedial literacies to redefine the competence of struggling adolescents. *Reading Online*, 6(7). Retrieved from http://www.readingonline.org/newliteracies/lit_index.asp?HREF=obrien2/

O'Brien, D. G., Springs, R., & Stith, D. (2001). Engaging at-risk students: Literacy learning in a high school literacy lab. In E. B. Moje & D. G. O'Brien (Eds.), *Constructions of literacy: Studies of teaching and learning in and out of secondary schools* (pp. 105–123). Mahwah, NJ: Lawrence Erlbaum Associates.

Olwig, K. F., & Harstrup, K. (Eds.). (1997). *Siting culture*. New York, NY: Routledge.

Pahl, K., & Rowsell, J. (2013). Artifactual literacies. In J. Larson & J. Marsh (Eds.), *The handbook of early childhood literacy* (2nd ed., pp. 263–278). London: Sage.

Pink, S. (2007). *Doing visual ethnography: Images, media and representation in research*. London: Sage.

Pink, S. (2009). *Doing sensory ethnography*. London: Sage.

Piot, C. (1999). *Remotely global: Village modernity in West Africa*. Chicago, IL: Chicago University Press.

Schmier, S. (2010). *Our space: Researching literacies and identities in and across classroom and online spaces* (Unpublished doctoral dissertation). Teachers College, Columbia University, New York, NY.

Siegel, M. (1984). *Reading as signification* (Unpublished doctoral dissertation). Indiana University, Bloomington, IN.

Siegel, M., & Panofsky, C. P. (2009). Designs for multimodality in literacy studies: Explorations in analysis. In K. M. Leander, D. W. Rowe, D. K. Dickinson, M. K. Hundley, R. T. Jimenez, & V. J. Risko (Eds.), *58th yearbook of the National Reading Conference* (pp. 99–111). Oak Creek, WI: NRC.

Skop, E., & Adams, P. C. (2009). Creating and inhabiting virtual places: Indian immigrants in cyberspace. *National Identities, 11*, 127–147. doi:10.1080/14608940902891161

Smith, A., & Kollock, P. (1999). *Communities in cyberspace*. London: Sage.

Soja, E. (2010). *Seeking spatial justice*. Minneapolis, MN: University of Minnesota Press.

Stern, S. (2007). Producing sites, exploring identities: Youth online authorship. In D. Buckingham (Ed.), *Youth, identity, and digital media* (pp. 95–117). Cambridge, MA: MIT Press.

Stornaiuolo, A., Hull, G., & Nelson, M. (2009). Mobile texts and migrant audiences: Rethinking literacy and assessment in a new media age. *Language Arts*, 86, 382–392.

Thomas, A. (2004). Digital literacies of the cybergirl. *E-Learning and Digital Media, 1*, 358–382. doi:10.2304/elea.2004.1.3.3

Thomas, A. (2005). Children online: Learning in a virtual community of practice. *E-Learning and Digital Media*, 2, 27–38.

Thomas, A. (2007). *Youth online: Identity and literacy in the digital age*. New York, NY: Peter Lang.

Turkle, S. (1984). *The second self: Computers and the human spirit*. New York, NY: Simon & Schuster.

Turkle, S. (1995). *Life on the screen: Identity in the age of the Internet*. New York, NY: Simon & Schuster.

Turkle, S. (2011). *Alone together: Why we expect more from technology and less from each other*. New York, NY: Basic Books.

Wellman, B., & Haythornewaite, C. (2002). *The Internet in everyday life*. Oxford: Wiley-Blackwell.

Wohlwend, K. E. (2008). Play as a literacy of possibilities: Expanding meanings in practices, materials, and spaces. *Language Arts*, 86, 127–136.

Wolcott, H. (1987). On ethnographic intent. In G. Spindler & L. Spindler (Eds.), *Interpretive ethnography of education: At home and abroad* (pp. 37–57). Hillsdale, NJ: Erlbaum.

· 4 ·

COMPOSING SOCIALLY

Abstract

This chapter demonstrates how mobile phone-based composing practices are highly social practices where composers draw on the words, ideas, arrangement or structure, and even tone of other composers. Not only is mobile phone-based composing highly social in that composers mobilize existing content for achieving specific effects, but it is also social in that users learn what is appropriate and intelligible through interacting in digital spaces. Furthermore, mobile phone-based composing practice is often shaped by its reception; that is to say, composers are interested in the feedback a post might receive and in some sense, the audience becomes a sort of co-author as audience feedback is taken into consideration as composers construct their texts. These ideas are explored through Bakhtin's theory of dialogism including heteroglossia, addressivity, and polyphony. Finally, this chapter shows how composing in digital environments necessitates consideration of audience attention or addressivity in a digital milieu where content is abundant and at times even overwhelming. This chapter also shows that in many ways, this kind of social composing is not all that different from school-sanctioned composing practices.

Keywords: Social composing, addressivity, audience, Bakhtin, schooled literacies, heteroglossia, Twitter, polyphony

This chapter aims to show how mobile phone-based composing practices are highly social practices drawing on the words, ideas, arrangement or structure, and even tone of others. Not only is mobile phone-based composing highly social in that composers mobilize existing content for achieving specific effects, but it is also social in that users learn what is appropriate and intelligible through interacting in digital spaces. Furthermore, mobile phone-based composing practice is often shaped by its reception; that is to say, composers are interested in the feedback a post might receive and in some sense, the audience becomes a sort of co-author as their feedback is taken into consideration as composers construct their texts. Finally, composing in digital environments necessitates consideration of audience attention in a digital milieu where content is abundant and at times even overwhelming. Composing is social in that composers pay attention to the findability of the texts they produce, employing particular techniques that take into account audiences' attention and audiences' needs. In many ways, this kind of social composing is not all that different, though, from school-sanctioned composing practices, as the close of this chapter will show.

Bakhtin's Dialogism and Mobile Phone-Based Composing

To understand the social nature of the practices engendered by the mobile phone, I draw on the work of Russian literary theorist Mikhail Bakhtin. Theoretically, the research presented in this book is rooted in a view of literacy as the process by which people learn how to use language through interacting with others and the world (Bakhtin, 1982, 1986). In doing so, we constantly take up others' language and signs and use them in our own communication. This idea, which Bakhtin termed "dialogism," grew out of his study of the novel (in his time a new textual form), which he saw as multivoiced and intertextual. Bakhtin's ideas have been applied generatively within the realm of language and literacy studies to examine the ways in which students' home and school environments contain voices in dialogue, how student identities are formed as a process involving linguistic forces in tension, and how recognizing the dialogic and multivocal nature of spaces for learning can inform classroom practice (e.g., Ball & Freedman, 2004; Nelson & Hull, 2008; Pleasants, 2008). I draw on Bakhtin's ideas to acknowledge the multitude of voices within and the social nature of digital composing practices using the mobile phone. Although multimodality and mobility are significant features of phone-based literacy practice, I rely on a Bakhtinian framework to examine the social and dialogic aspects of the practices under consideration in particular. Specifically, I take up Bakhtin's concepts of addressivity, heteroglossia, and internally persuasive and authoritative discourse as a conceptual framework in thinking about the ways in which composing with mobile phones is a highly social rather than individual practice.

Addressivity

According to Bakhtin (1982), any signifying practice represents a two-way engagement with a social world. Although every utterance is present "at a particular historical moment in a socially-specific environment" (p. 276), it at once both speaks back all utterances that have come before it and anticipates a response in the future (a characteristic he called "addressivity," to suggest that it is directed toward an other). A text, then, represents not something devoid of influence or history, but a link in a larger chain of communication (Bakhtin, 1982) and thus production and reception are mutually constitutive. Rather than being completely confined as communicators to the conventions of existing language systems, however, Bakhtin (1982) notes that

> Every utterance participates in the unitary language (in its centripetal forces and tendencies) and at the same time partakes of the social and historical heteroglossia (the centrifugal, stratifying forces). (p. 272)

Heteroglossia

Bakhtin (1982) used the term "heteroglossia" (in Russian, literally "different-speech-ness") to describe the ways language is at once populated with one's authorial intent and the words and authorial intent of others. Bakhtin (1982) wrote that this double-voiced nature of language was "another's speech in another's language, serving to express authorial intentions but in a refracted way" (p. 324). This idea of language as heteroglossic acknowledges the hybridizations, the

> mixture of two social languages within the limits of a single utterance, an encounter, within the arena of an utterance, between two different linguistic consciousnesses,

> separated from one another by an epoch, by social differentiation or by some other factor. (p. 358)

Such hybridization is common in online communicative practice especially in an age of remix (Lankshear & Knobel, 2008), reblogging or text curation (Mihailidis & Cohen, 2013), and cut-and-paste (Roth, 1999), all practices that are heteroglossic writ large (Nedeljkov, 2011). In his discussion of heteroglossia, Bakhtin (1982) delineated the concepts of "authoritative" discourse and "internally persuasive" discourse, describing the two as enmeshed in an ongoing struggle. Authoritative discourse is

> located in a distanced zone, organically connected with a past that is felt to be hierarchically higher. It is, so to speak, the word of the fathers. Its authority was already acknowledged in the past. It is a prior discourse. (Bakhtin, 1982, p. 342)

Authoritative discourse, then, is language that is already in existence that is attributed some kind of authority or power. It is privileged language that is thought of as official or authentic.

Bakhtin (1982) put forth internally persuasive discourse as in opposition to authoritative discourse, a "retelling in one's own words" (p. 341), imaginative discourse that is "tightly interwoven with one's own word" (Bakhtin, 1982, p. 345). For Bakhtin, people have agency as communicators, but are inextricable from their context, linguistically and historically. In other words, through his distinction between authoritative and internally persuasive discourse, Bakhtin describes the hybrid nature of language: a person's words are ever populated with historical meanings, even as she or he contributes authorial intent. To use youth writing online as an example, authoritative discourse might be the standard American English taught in school or the discursive conventions associated with a particular platform (e.g., hashtagging on Instagram or long-form writing on a blogging platform). Internally persuasive discourse might be the inner, private voice of the individual youth composing on a particular platform. A consideration of both of these voices, the authoritative and the internally persuasive, is necessary to better understand the social nature of the digital literacy practices of youth.

Composing Socially

Dominant approaches to literacy frame composing as an individual accomplishment. Youth are assessed alone, and texts are assumed to have been the "original" work of the individual. However, composing online (and certainly communicating through mobile phones) is highly social. By situating literacy in the individual as an act of mind, the influence of the social milieu is neglected. Consumers of networked digital media have the ability to quickly and easily interact with authors through comment boxes, social-networking buttons, or email, making digital composition more potentially dialogical than print composition. The properties of networked, multimodal, potentially interactive and global digital composing spaces also allow for remixing and recirculating, practices that challenge the notion of solo authorship. In digital composing spaces, it is easy to "cut," "paste," and "remix" a person's work. A digital text posted online can be easily remixed, remediated, or transmediated (Siegel, 1995) by others online by consumers who "view textual property not as the exclusive domain of textual producers but as open to repossession by textual consumers" (Jenkins, 2006, p. 60). Composers online are engaged, consciously or not, in a struggle for "possession of the text and for control over its meanings" (Jenkins, 2013, p. 24). The rising popularity of GIFs and memes, which take snippets from movies, television shows, and other popular culture artifacts and make those images or short clips available for communicators to use within their own individual conversations, is a good example of this struggle, and illustrates the highly social nature of composing in digital environments.

According to Bakhtin (1986), using language is about more than control of words, clauses, and punctuation. Knowing how to use language is not simply a matter of adhering to the authoritative rules of grammar handbooks and dictionaries. It is about knowing how to participate in ways that conform to social norms (to some extent) to be understood.

The data collected in this study demonstrated they ways in which the approaches to composing in the digital space for these participants were socially constructed. For example, their practice of digital curation represented social, not solo, composing. This was a practice in which they spoke through the voices of others, and this afforded them a scaffolded method of interacting in the digital space. This social method of composing also influenced the way they formulated their own original utterances in those spaces. In addition, they used feedback in those spaces to formulate their own utterances. They learned how to communicate in the digital space through social interaction. They used digital media to mediate their experiences as adolescents working to construct their identities (Subrahmanyam & Smahel, 2011). For example, Eli described the process of creating his social media pages to communicate with friends, and his Instagram and Twitter represented heteroglossic (Bakhtin, 1982) digital collections of artifacts, photographs, and sentiments, which were located at the nexus of his social connections.

The participants determined what was applicable, socially fitting, and what would be understood or appeal to his audience, who were not physically or temporally present, through the feedback they received in their digital spaces that allowed for such interactivity. Many times, they would determine this through the responses they got to their postings, and a lack of response would almost certainly lead to their deleting a particular posting. I could determine that the language they used mirrored that of their friends through observing their Web spheres. Many of the same phrases, syntax, and other conventions were observable both in the participants' postings and the commentary from other friends on those posts, for example. They learned about more widely-shared conventions on the social media platforms where they composed, such as #ManCrush Monday or #MCM, where people post photos on Monday of male crushes, #WomenCrushWednesday or #WCW which applies the same concept of #MCM to Wednesday, and #ThrowbackThursday, where people post older photographs to celebrate a happy memory. Sometimes they would remix these conventions, demonstrating Bakhtin's concept of internally persuasive discourse. For example, John posted his own versions of these weekday categories with his own hashtags such as "#TurnUpThursday" where he posted a photo of himself "turning up" or partying. The youth engaged in complex practices as they worked to meet social norms for written and spoken communication online while at the same time expressing their own unique identities.

When exploring Eli's Web sphere, by following the links to pages of those with whom he interacted on Twitter, I could see that these youth were also composing tweets that were primarily affective in nature. Figure 4.1 shows the last seven open tweets posted by four users with whom he interacted.

User		Last seven tweets
Friend 1	1	I miss my freakin boyfriend
	2	I'm sleepy again.
	3	I wish my babe was up ….
	4	I hate when niggas say "so where yo man at?" Bitch til I tell you I got a man, DON'T assume I do …. But shit, he right here. Aha!
	5	I woke up & my parents wasn't here …. Shit for all this I could have stayed with Tony!!! Tf.

User		Last seven tweets
	6	I shouldn't have went to sleep so early
	7	Why am I up?
Friend 2	1	I wonder if mommy wants to go out to eat.
	2	It's been like a month and some change and that's not gon work if we tryna make things happen.
	3	Oomf needa come see me.
	4	I'm soooo bored.
	5	I miss kissin on you, you know I miss kissin on you.
	6	Nobody understands the situation I'm in.
	7	When I'm single I do as I please, but as soon as I get locked up, I'm the most committed, faithful, down to ride female ever.
Friend 3	1	G NiGht
	2	Ha, Time Tuh Man Dwn …
	3	Only Weakness … Muh Shawty
	4	Yeah I'm On Sum Soft Shxt ToniGht …
	5	Ain't Shxt I Guess :/
	6	She On Dat Heart 6reakinG Shxt …
	7	Fckers PlayinG Games … Ahh I See
Friend 4	1	Clarence holding me up ! I need to go get my outfit for tomorrow
	2	Don't feel like this bullshit
	3	I gotta clean the whole house <
	4	Coolin with my babies tomorrow >
	5	Im ready for tomorrow >>>
	6	I swear I be wanting to go back to my old ways
	7	Bitches is fake * shrugs *

Figure 4.1: Tweets from Eli's Web Sphere.
Source: Author.

Eli's cultural models around microblogging were socially constructed. He learned through dialogic interaction with his social world how to communicate in a way that will allow him to be understood (Bakhtin, 1982, 1986). Like those he interacted with in his Web sphere, the content of his tweets were affective, and he used oral-language structures. Because he was not afforded direct instruction in how to use Twitter to communicate, it makes sense that he would turn to the practice of those around him in his Web sphere, those to whom he was connected, to learn how to interact there. It was through engagement that he learned how to communicate in the digital space in a way that would be understood by others and that adhered to the conventions of that space (while contributing his own intention).

Like Eli, the majority of John's original (i.e., not retweeted) tweets were related to emotions:

> Might be the happiest day if my life
> On my way to school … swear I hate ths sht
> whats the point
> there's no place like home
> you make me feel like I can fly
> feels good to be bakk home
> today oomf [one of my followers] piss me off widd da sht she was doin man
> school been funny af today
> just uh beautiful day

However, his tweets differed from Eli's in that he also posted pithy, "spreadable" (Jenkins, Ford, & Green, 2013) messages. According to Jenkins, media content is spreadable if it is shown to be valuable and meaningful through "shares" (in the context of Twitter, the share is called the retweet). Interestingly, it was only after I saw the kinds of messages that John was retweeting that I recognized in turn the spreadable nature of John's tweets. The messages John was retweeting were succinct, quotable messages in the style of aphorisms, adages, maxims, or truisms:

> Don't change who u r cause your past just be wise
> Thankin god for another 24
> Ion apologize for what I say.
> People That Say; "I Dont Care." Are The People That Care The Most.
> don't evah be sorry for telling the truth
> Ramen noodles always come in handy when it aint shit in the house to eat …
> I'll stand alone before I follow behind another.
> People gone drag your name in all type Of dirt but actions always show the real you

John produced many of these quotable, spreadable maxims himself, in some cases these represented reconfigurations of rap lyrics and others of these tweets were his of his own invention:

> The more I give the more they tryna take
> don't ever let uh nigga think yu need him.
> I was taught ta say FUCK uh friend be about yo business.
> Fck dem fake friends, way yo real friends
> a #MILLION words is nothing compared to showing someone
> sometimes I have them thoughts like, um too real fo this sht

John draws on the conventions and even whole utterances within the social spaces where he composes heteroglossically in crafting his own texts. At once, he evokes the cultural meanings encoded in the texts and spaces and creates new ideas through recontextualizing these cultural bits and pieces. This makes his communication comprehensible to his followers and others who may come across it, but it also serves as a marker of identity and group belonging and is a means of building social capital. Through taking up the texts and practices in the spaces where they interact, both Eli and John's communication are local and translocal as ideas, conventions, and cultural messages travel across digital and physical spaces through composers' practices.

The participants relied heavily on the texts of others in their digital composing practices, too, in that their feeds integrated whole utterances from other users' feeds. Eli's Twitter feed, for example, was not a single-authored text; instead, it was a collection of texts from his various social connections. It was polyphonic (Bakhtin, 1982). This kind of composing practice is social in ways that are highly visible. Eli's Twitter feed was not a single-authored text; instead, it was a collection of texts from his various social connections. In this way, it was not monologic but instead it was polyphonic (Bakhtin, 1982). As a representation of the self and a cohesive text, the Twitter feed did contain a dominant authorial voice: Eli's voice. While it was heteroglossic (Bakhtin, 1982) writ large, there was some verbal unity. There is one authoritative voice that works to bring the various other voices together (e.g., friends, rappers, celebrities, and other Twitter users). With a number of tweets originally authored by others, Eli's Twitter feed was certainly polyphonic, but as he used the words of others to express himself, he imbued them with new meaning in a heteroglossic way. For example, in tweeting the lyrics to a song (for example, "I wish a nigga would like a kitchen cabinet" from a song by the rapper 2Chainz), he took what had a particular meaning for 2Chainz, and then imbued it with new meaning in the context of his Twitter page (e.g., "I appreciate this lyric"). He especially liked to curate posts from others that contained double meanings and innuendo or "signifying" (Lee, 1995, p. 380). In this example, the rapper 2Chainz is playing on the word "would" by comparing it to the wood from which a kitchen cabinet might be constructed. 2Chainz is known for such absurd wordplay, remixing the African American linguistic tradition of signifying, and Eli demonstrates that he "gets" what sets this rapper apart by choosing to focus on the more outlandish lines in his songs. The text of the Twitter feed was one that represented Eli's life and values, and as a reader, we come to know about his life and values through conversations with friends, retweeted material that reflects what he appreciates and cares about, photographs that capture important people objects, and images of his adorned body.

The participants' digital composing practices were also social beyond what is visible to a casual observer. I came to recognize the collaborative nature of online composing early on in the data collection phase of this research. I began to notice that the participants were deleting posts almost as frequently as they were creating new posts. Interviews helped to clarify that they were using feedback in the space to make decisions about revising their texts. One day, I was carrying out participant observation on Instagram when Eli posted a compromising photo of himself. After feedback from several friends indicating shock that he would post the picture, he deleted the picture. In this way, he used feedback from others as he socially constructed his text. He considered what others thought (addressivity) as he formed his utterance. Similarly, the participants reported that they would delete postings on Instagram, for example, that failed to garner enough "likes."

Bakhtin posits that discourse is historically and socially produced and is also designed with what will be understood by the addressee or superaddressee in mind. Inherent to Bakhtin's (1986) theory of dialogism is the idea that all discourse contains aspects that imply anticipation or response in word choice and grammatical and syntactic features. Central to dialogism is Bakhtin's (1986) theory of addressivity, in which he maintains that a writer is influenced by an imagined or anticipated audience reaction. Bakhtin (1986) contended that

> speakers always shape an utterance not only according to the object of the discourse (what they're talking about) and their immediate addressee (whom they're speaking to), but also according to the particular image in which they model the belief they will be understood, a belief that is a priori of all speech. Thus, each speaker authors an utterance not only with an audience-addressee, but a superaddressee in mind. (p. xvii)

For example, when youth post selfies to online sites hoping to garner a particular kind of feedback they demonstrate addressivity. According to Bakhtin (1986), using language is about more than control of words, clauses, and punctuation. Knowing how to use language is not simply a matter of adhering to the authoritative rules of grammar handbooks and dictionaries. It is about knowing how to participate in ways that conform to social norms (to some extent) to be understood.

The participants' practices demonstrated addressivity, yes, but in this highly interactive context, the participants were also able to use what was often instant audience feedback in their composing process. They did not have to rely just on their imagination to anticipate an audience response; they often received responses from their audiences fairly immediately. In this way, the texts they created were socially constructed to some extent. That is to say, audience feedback shaped the production of their texts in concrete and observable ways. When the participants deleted texts that garnered little, no, or negative feedback, the production and reception of those texts were shown to be mutually constitutive. Rather than an individual accomplishment, this kind of literacy practice represented a highly social form of composing, one that I would argue was not possible before the age of ubiquitous computing and networked composing with applications supporting this kind of interactivity.

Composing as Identity Construction

The like button or like option, a feature built into many social networking platforms, is a means of providing feedback to users about the content they post. The like button can be used in a variety of ways; it might be used to show appreciation for content, to show acknowledgment, as a show of support, or some other demonstration. The number of users who choose to like a particular piece of content is usually displayed under the content and can act as a quantitative measure of the popularity of that item. On Instagram, this feature appears as a heart that a user can tap to express a "like" and on Twitter, in order to show appreciation for posted content, one could retweet or favorite it. Since this study was conducted, Twitter renamed the favoriting feature "like" and replaced the favoriting icon with a heart icon like the one that appears on Instagram.

On some platforms, liking content as a form of acknowledgement became somewhat controversial. For example, on Facebook, the only option to show that one saw a particular posting short of commenting was to like the post.

However, there was a kind of cognitive dissidence that resulted when users posted about unfortunate news and other users pressed Facebook's thumbs up icon. The result was that some users felt that their need to acknowledge the content somehow looked like schadenfreude. As a result, in 2016, Facebook introduced "reactions," a redesign of the like button that includes a wider array of expressions including "wow," "angry," "haha," "sad" and love. Users can choose from five emoji to express these reactions. While users could have commented using words or linguistic text to express more nuanced reactions to content, since most users access Facebook using mobile phones, the company decided to offer the tappable emoji.

If we believe that youth use social networking platforms as a means of gaining information about themselves and about the world as a process of identity construction, it follows that the like button is a mechanism by which youth can gain information about themselves. As Bakhtin (1984) purports, it is through dialogic interaction with the world that the social subject comes into being. It is through "mutual reflection and mutual acceptance" (p. 287) that youth come to know themselves. The self is inextricable from the social world. As Burkitt and Sullivan (2009) explain, the human social being will "be aware of the many different definitions of ourselves from the worlds and actions of others, having to bear the fact that—while fixed and defined to some degree—we are always unfinalized and open to constant redefinition. The ability to achieve [a unified social self] rests on the ability to create a unified voice for ourselves, strong enough to speak with some authority in dialogue with others and resistant enough to withstand being torn apart" (p. 565). In other words, centrifugal forces of the voices of others work to stratify our conceptions of our selves while the centripetal force that is the unity of our being works to unify or centralize the self. Bakhtin borrowed these terms, centripetal and centrifugal, from physics to demonstrate the ways in which this dialogic interaction allows the social self to be in flux rather than stable.

Posting online to spaces where youth can receive feedback in the form of likes or other reactions is a way for them to see themselves in a divided way, through another's eyes. These postings, by their very nature (that is to say, posted to social networks in an addressive way in a context that allows for networked publication within a platform that offers mechanisms for feedback) invite such feedback. The action of posting does not end with the posting. It does not exist on any particular temporal scale, as a young person can reflect on the amount or type of feedback that a posting garnered well into the future and take away from that information about their identity or other social information.

Social Writing and Audience

Online using their phones, the participants, engaged in social writing practices like digital curation and meme-sharing, which were highly visible to their followers. They had, however, the option to control the flow of their communication through limiting the audience for their posts either through using controls built into the social media platforms they used or through sending direct messages or DMs, a tool for messaging someone privately. Through using audience-limiting practices, they could target their communication for particular audiences and have more intimate conversations.

Audience-limiting practices have become so common that certain conventions have emerged that are a kind of unwritten rule. The phrase "slide into the DMs" has come to be synonymous with flirting through social media, and rapper Yo Gotti's hit "It Goes Down in the DMs," a song describing the importance of direct messaging as a form of modern communication, won a Black Entertainment Television award in 2016 because it resonated with so many in Generation Y. In Yo Gotti's DM anthem, he sings, "Don't you hate when you get screenshotted? (petty)/Bitch that DM weren't for everybody (rules)" referring to the sacredness of the DM as a means of communicating privately. Screenshotting (in order to save or share) a private message is a violation of the unwritten rules of DM etiquette because it means a loss of control for composers looking to limit the audiences for their digital texts.

Another means of limiting the audience for mobile-phone based communication is through a subtweet, as described in Chapter 3. Through imprecise language, metaphor, or other means of masking ones' meaning, the audience can be limited and in some ways privatized. Youth have done this for ages through secret languages and inside jokes.

Eli had a sense of what was acceptable discourse in the public space and what should be relegated to private spaces, like the private messaging features on Twitter or a direct phone call:

> Eli: Like, I don't really like having conversations with more than one person.
>
> Julie: It seems like some people are talking directly to each other but they put it like out where everybody can see it. What's the point of it?

Eli: Like when you want to have a conversation, like, you don't want nobody to know who you talkin' to.
Julie: Oh! Oh, oh, oh, oh. But that's not even what I'm talkin' about. Like …
Eli: We havin' a conversation on Twitter but it's just statuses?
Julie: Yeah.
Eli: Like, "Hey, what's up? @whatever."
Julie: Right, yeah.
Eli: And then they be talkin' about it for … that stuff annoys me I don't know why.
Laughter
Eli: Like, just call each other. Why are you wasting my page status?

Eli had a clear notion about how one should navigate the various spaces to communicate with the intended audience. He used provocative language on Twitter that was markedly different from the language he used in the classroom space (though not different from the hallway space). He used derivative of the pejorative term, "nigger," historically used to oppress, as a term of endearment (Mahiri, 1998) and as a form of (neutral-intent) address. Though qualitatively different from its historical use, his use was not what Smitherman (2001) called "script flippin" because Eli is outside African American culture and thus cannot reclaim the term (Mahiri, 1998). The word was frequently evoked in the music he listened to and by his friends, and I saw his use of it as a way to align himself with African American culture. He used language in a way that showed heightened awareness of how the words he used operated in the world.

Strategies for Reaching Audiences

Part of composing socially is thinking about reaching your intended audience. In an age where digital content is abundant, composers can employ certain techniques to make their content visible and searchable to target audiences. Search Engine Optimization (SEO), or the act of tailoring the content of a website so that it is more visible within a set of search results, is employed by novice producers of online content as well as savvier Internet marketers. For example, one might consider commonly searched keywords as well as synonyms for those keywords and edit site content to include those words. This strategy shows heightened consideration of audiences and anticipates in a way reminiscent of Bakhtin's addressivity an audiences' response to a particular communicative need.

Hashtags, or words or phases preceded by the pound or number sign, are used to label content online in order to make it searchable. For example, the youth in this study would often use hashtags to label a post with a hashtag identifying the theme or some other aspect of the content of the posting. The use of such metadata is another example of a technique used to anticipate audience needs.

Timing and automation are other strategies that composers of digital content employ that take into account their audience. Algorithms can determine peak days and times for audience engagement, and composers can decide to publish their posts during those times in order to ensure that their texts are consumed by their audiences. Some digital content creators use software to automate the dissemination of their content at peak times for engagement. The participants in this study knew that their audiences had generally the same peak times for engagement that they did (in the morning before school, after school let out in the afternoon, after the evening dinner hour) since they all had very similar schedules as high school students.

Social Composing at School

Such social composing is not exactly new, or a function of the mobile platforms for composing. In carrying out observations in the participants' school, I noted the ways in which they were asked to engage in social composing practices as part of their official literacy curriculum. For example, students were asked by Ms. Hamilton to engage in curating activities as part of their composing practice as they wrote argumentative or persuasive papers, as my field notes illustrate:

May 14, 2013: The students are working on argumentative papers.
Ms. Hamilton: Brainstorm ideas for your argumentative paper. You should have written your thesis. Your thesis is your opinion plus two reasons.
(On Promethean board):
Name:
Date:
Argumentative Organizer
Thesis (Your opinion plus two reasons)
Should gun laws be stricter?
Ms. Harrison asks the students to give her an example of one of their thesis statements.

Student: Wear uniform would prevent from bullying.

Ms. Harrison: (Helping him to phrase it) "School uniforms are necessary because … " and then move on with the reasons.

She asks them to raise their hands if they have a thesis statement.

Ms. Harrison: Those with their hands up, find someone who needs help and help them.

Ms. Harrison adds to the outline on the screen as the students converse.

(On the Promethean board):

Topic Sentence: One reason that there should be stricter gun laws is that innocent people are often killed.

Two students offer, "Sandy Hook!" "Sandy Hook!"

Ms. Harrison asks for examples.

Student: Teen suicide.

Ms. Harrison: What are your two reasons?

Student: Friends, and anxiety.

Ms. Harrison: Ok, when I come around everyone should have one topic sentence.

John is goofing off and singing, but Ms. Harrison ignores him. She comes around to help anyone not finished with their topic sentences.

Ms. Harrison: Now give me two pieces of support or two examples.

They begin to contribute.

Sandy Hook! Colorado movie theater incident!

Then one student stands from her seat and says, "OH! Last night, on 48 hours … "

Ms. Harrison: Wait, was it a true story?

Student: No …

Ms. Harrison: Well I don't want to hear it then.

Ms. Harrison comes around to look at their papers to check out the students' examples. They're busily writing their outlines.

Ms. Harrison: If you don't have a good outline, you're going to struggle to write your paper.

Student: We're already learning this in another class.

Student: Miss Harrison! I got somethin I think.

She comes around and reads and critiques their work.

Ms. Harrison: You have three minutes to write your text.

Ms. Harrison helps a student to come up with ideas for her writing. The timer that Ms. Harrison has set to denote the allocated time has passed goes off.

Ms. Harrison: Ok, if you are struggling, which I think you are, I'll give you some time to do research and try to find support. You can only use [name of state-sponsored virtual library, available through a website] or Opposing Viewpoints (online resource licensed to the school from Cengage Learning) because I know those are credible sources.

The students begin to use these websites to inform their brainstorms.

Ms. Harrison: Remember, you need one quote. If you do not have your iPad, when I finish talking, you're going to get a computer. You might find something new. It's ok to change your thesis. That's why they call it a working thesis. You're not on websites, you're not on Google, you're on [state-sponsored virtual library]. Click on "A-Z." Remember what I told you about sources. If it's a direct quote, use quotation marks. If it's a paraphrase, you still have to tell me where you got the information from. Try to stick to that same section where it says Opposing Viewpoints. Try not to go to the Reference Desk.

As they work quietly, Eli gets called to the in-school suspension office. The students log on, surf around, read, and look for information on their topics. It is the quietest it has been all class. They seem focused and engaged. Ms. Harrison gives them 15 minutes to surf these pages and find information they can plug into their outlines. As I observe the students, I see the topics range from abortion, to domestic violence, to foster care. They are being asked to formulate their own opinions, but to use the website to inform their perspectives.

Ms. Harrison: Alright, I'm gonna allow you the rest of the time to research, but I'm going to tell you one thing. You need a counter argument. You can't just write a whole paper and never address the other side. Find the other side. Somewhere on your paper, label it counterargument. Don't forget to write where you got it from.

As this excerpt illustrates, the students were learning how to compose socially at every point in the writing process. In the brainstorming phase, as they came up with topics to address, they were given time and explicit instruction to help each other come up with topics on which to write, coming at their idea by way of social interaction. As they came up with ideas and early thesis statements, the teacher acted as an audience, providing the students with feedback. Next, they were directed to use websites (sites which themselves were likely constructed socially) to inform their argument as they began to put meat on the bones of their early draft. They were asked to draw on prior instruction around how to integrate the voices of other writers though summarizing, paraphrasing, and quoting. They were asked to anticipate readers' reactions by addressing the opposition through integrating a counterargument. Through drawing on the voices of others, they were able to jointly construct knowledge and to create a more robust and contextualized piece of writing.

Conclusion

In this chapter, I aimed to show the ways in which composing in digital spaces with and through mobile phones is a highly social writing practice. That is to say, audiences have perhaps as much control over the content and publication of a digital text as composers have. Composers must not only adhere to social norms in different digital spaces, but they must be aware of how language functions in order to know how they will be "read" in online space. They must take into account audiences' limited attention and employ techniques that will increase or limit their audiences.

Not only does the audience online, through feedback offered in the form of comments, likes, views, or even silence, shape the kinds of literacy practices and texts young people engage in and produce, the engagement of others in the spaces shape what youth produce both structurally and in terms of content as they learn to compose through social interaction in the space. And not only do audiences shape texts as youth write with, for, and in many ways through others, they also shape the composers of those texts, who learn about their identities and their social worlds through interaction in digital spaces.

The social nature of these mobile phone-based composing practices aren't new, however. As the close of this chapter shows, dialogic composing practices are constitutive of many traditional school-based literacy practices. Persuasive papers and research papers, which have a long history and which seem to have potential staying power when it comes to English language arts curriculum, are highly dialogic and social.

However, it would benefit students who compose with mobile phones to be recognized for having skill in this arena—they are composers with heighted and nuanced senses of audience awareness. Ideally, these insights could be applied to support classroom practice that takes into account the rich background in social writing that students who compose with mobile phones have in their toolkits.

References

Bakhtin, M. (1982). *The dialogic imagination*. Austin, TX: University of Texas Press.

Bakhtin, M. (1984). *Problems of Dostoevsky's poetics* (C. Emerson, Trans.). Minneapolis, MN: University of Minnesota Press.

Bakhtin, M. (1986). *Speech genres and other late essays* (V. W. McGee, Trans.). Austin, TX: University of Texas Press.

Ball, A. F., & Freedman, S. W. (2004). *Bakhtinian perspectives on language, literacy, and learning*. New York, NY: Cambridge University Press.

Burkitt, I., & Sullivan, P. (2009). Embodied ideas and divided selves: Revisiting Laing via Bakhtin. *British Journal of Social Psychology*, 48(3), 563–577.

Jenkins, H. (2006). *Convergence culture: Where old and new media collide*. New York, NY: New York University Press.

Jenkins, H. (2013). *Textual poachers: Television, fans, and participatory culture*. New York, NY: Routledge.

Jenkins, H., Ford, S., & Green, J. (2013). *Spreadable media: Creating value and meaning in a networked culture*. New York, NY: New York University Press.

Lankshear, C., & Knobel, M. (Eds.). (2008). *Digital literacies: Concepts, policies, and practices*. New York, NY: Peter Lang.

Lee, C. D. (1995). A culturally based cognitive apprenticeship: Teaching African American high school students skills in literary interpretation. *Reading Research Quarterly, 30*(4), 608–631.

Mahiri, J. (1998). Streets to schools: African American youth culture and the classroom. *The Clearing House, 71*(6), 335–338.

Mihailidis, P., & Cohen, J. (2013, Spring). Exploring curation as a core digital and media literacy competency. *Journal of Interactive Media in Education, 2*, Spring.

Nedeljkov, N. (2011). Liquid identity: Cultural exchange between the reader and the text. In B. Frymer, J. Broughton, & M. Carlin (Eds.), *Cultural studies, education, and youth: Beyond schools* (pp. 53–71). Plymouth: Lexington.

Nelson, M. E., & Hull, G. A. (2008). *Self-presentation through multimedia: A Bakhtinian perspective on digital storytelling*. In K. Lundby (Ed.), *Digital storytelling, mediatized stories: Self-representations in new media*. New York, NY: Peter Lang.

Pleasants, H. (2008). Negotiating identity projects: Exploring the digital storytelling experiences of three African American girls. In M. L. Hill, L. Vasudevan (Eds.), *Media, learning, and sites of possibility* (pp. 205–233). New York, NY: Peter Lang.

Roth, L. (1999). Educating the cut and paste generation. *Library Journal, 124*(18), 42–44.

Siegel, M. (1995). More than words: The generative power of transmediation for learning. *Canadian Journal of Education, 20*, 455–475.

Smitherman, G. (2001). *Talkin that talk: Language, culture, and education in African America*. London: Routledge

Subrahmanyam, K., & Smahel, D. (2011). *Digital youth: The role of media in final development*. New York, NY: Springer.

· 5 ·

CHALLENGING THEORIES OF DESIGN

Abstract

This chapter challenges design theory as a lens for examining mobile phone-based literacy practices through examples of the ways that the affordances and architecture of the mobile phone and the engendered practices challenge previously-dominant views of composing as planned, designed, and focused on an end-product. Instead, this chapter argues, mobile phone-based composing practices show composition as part of ongoing literacy practice engaged by youth as part of their larger social aims. Much of mobile phone-based digital composing happens on-the-go, in the in-between spaces of our lives, and, as the data show, may be more precipitous than previously theorized. In addition, mobile digital composing is highly dialogic and responds to existing voices and existing texts. Rather than springing forth spontaneously without origin or genealogy, so much of mobile phone-based composing is reactive and answers or otherwise responds to existing communication. This chapter begins by tracing the origins of the application of design theory to work in the field of digital literacies, and then explores how this idea of multimodal composing as a process of design materializes in some of the early digital literacies literature. These ideas are explored through the data collected in this study to demonstrate the ways in which the participants' writing practices challenge theories of mobile phone-based composing practices as design-based practices with attention to texts created on apps such as Snapchat and Instagram, which challenge ideas about revision and the temporal scales involved with composing.

Keywords: Design, design theory, mobility, mobile literacies, multimodality, dialogism, temporal scales, revision, Snapchat, Instagram

This chapter presents a broad analysis of mobile phone-based composing practices to challenge widely held notions about digital composing as a process of design. As I have demonstrated in other chapters of this book, much of mobile phone-based digital composing happens on-the-go, in the in-between spaces of our lives, and, as the data collected by way of this research show, may be more precipitous than previously theorized. In addition, mobile digital composing is highly dialogic and responds to existing voices and existing texts. Rather than springing forth spontaneously without origin or genealogy, so much of mobile phone-based composing is reactive and answers or otherwise responds to existing communication. This chapter begins by tracing the origins of the application of design theory to work in the field of digital literacies, and then explores how this idea of multimodal composing as a process of design materializes in some of the early digital literacies literature. I then flesh out these ideas using the data collected in this study to demonstrate the ways in which the participants' composing practices challenge theories of mobile phone-based composing as design-based.

Origins of the Application of Design Theory to Digital Literacy Practice

Because digital composing as a widespread practice is a relatively recent phenomenon, those who study so-called "new literacies" represent disparate fields, some residing in the fields of literacy education; some in English composition, rhetoric departments, and digital humanities; some in the fields of mathematics, science, and technology; some in communications studies; and still others in anthropology and cultural studies. Despite this variance, much of the work concerning multimodal composing and what it means to be a literacy educator in "new times" (Alvermann & Hagood, 2000; Gee, 2000a, 2000b; Luke, 1998; Street, 2003) amidst shrinking conceptions of literacy (Siegel, 2006) in the current assessment-saturated (Kontovourki, 2012) literacy-education milieu examines computer-mediated composing using the lens of design theory.

The approach to multimodal composing as a process of design can be traced back, in part, to early work in the field of composition studies. In this body of research from the 1980s, researchers sought to describe the ways young people approached a writing task. Most of these studies were conducted in schooled literacy contexts where students responded to an assigned writing prompt, and composing was framed as a "problem-solving" situation (Atlas, 1979; Berkenkotter, 1981; Cohen & Riel, 1989; Flower & Hayes, 1980a, 1980b; Fountaine, 1988; Rafoth, 1985; Reinking, 1998; Roen & Willey, 1988; Rubin & Rafoth, 1986). These studies were rooted in the assumption that composing requires the writer construct a mental model of the rhetorical situation so they may then surmount it through the act of composition. These studies were rooted in a constructivist epistemology that assumed that people construct mental models and that research could reconstruct those mental models to examine how people process information. Using think aloud protocols, researchers attempted to reconstruct a model of the rhetorical problem as they believed the writer conceptualized it. The model might include, as in one study by Flower and Hayes (1980a), reader, persona or voice, meaning, and text produced. They conjectured that writers store "problem representations" for handling conventional tasks like writing a thank-you card, but in new writing situations, good writers work to build a "unique representation" (Flower & Hayes, 1980a, p. 25). As is apparent through their design, these studies took a cognitive perspective rooted in schema theory that framed writing as an act of cognition and the mind as individual. This focus on the writer's composing processes challenged the then dominant focus on the written product in composition research. Yet, by situating literacy in the individual as an act of mind, this perspective ignores the influence of the social milieu of which the individual is part.

Landmark studies by Scribner and Cole (1981), Heath (1983), and Street (1984) laid the groundwork for the shift from an "autonomous" view of literacy as a set of cognitive skills to social theories of literacy that illuminate the ways literacy is embedded in social contexts. Contemporary literacy studies emphasizing the social practices of literacy, often called the "New Literacy Studies" (e.g., Barton, 1994; Fairclough, 1989, 1992; Gee, 1996; Heath, 1983; Lankshear, Gee, Knobel, & Searle, 1997; Street, 1984, 1995, 1999, 2003), move away from an earlier view of literacy as based in cognitive processes.

By the late 1980s, as the personal computer appeared, computers became more commonplace, and researchers turned their attention to computers' effects on composition. In one of the first studies of computer-mediated composing, Cohen and Riel (1989), following from the traditions of composition studies, examined the effect of "functional" writing environments—those that have purposes and authentic audiences, in contrast to the pseudotransactional (Petraglia, 1995) kinds of writing usually done in schools—and "distant audiences" by early email on students' writing. These researchers used an ecological-validity approach that attempted to replicate the contexts for writing in the world outside the classroom (i.e., "real"-world writing) by providing an audience for student writing through networked computers. Subsequent research on computer-mediated composing took a similar approach, applying composition studies-style approaches to researching computer-mediated composing processes. Working from an expanded view of modalities—examining use of images, color, size, fonts, and layout—researchers examined multimodal texts and used interviews with participants to learn more about how they make choices as composers (e.g., Beichner, 1994; Garthwait, 2007; Karchmer-Klein, 2007).

As literacy research took a "social turn," researchers of the technology-mediated literacy practices of youth began to shift from a focus on the process and the products of composing to the social contexts that gave rise to such practices as well as the social uses and meanings of those practices. It also inspired a reconceptualization of the digital text as researchers recognized the digital text as not only an object, but as a discursive space. In other words, the digital text was not only an aesthetic artifact, but also a space for connection and interaction. Accordingly, digital literacies researchers began to examine both the semiotics but also the social elements inherent to digital texts, both in terms of how they act as social spaces but also how they function in the wider scope of the participants' social worlds.

For example, Lewis and Fabos (2005) examined the social functions and the semiotics of IM through an examination of the ways seven young people put it to use. They carried out research that focused on language use, social networks, and surveillance in IM conversations. The researchers found that IM was more than simple conversational exchange; instead, they described how participants used IM in a performative way, tailoring their writing depending on their audience, shifting among tones, styles, subject matter, and digital grammars, as they interacted with different audiences. They also noted the ways participants addressed some multimodal aspects of the communication, using "nonlinguistic visual elements to supplement language in ways that served their needs" (Lewis & Fabos, 2005, p. 483). These included ellipses, codes such as (*), font, color, size, and emoticons. These nonlinguistic visual elements were tools or "one more way to express and engage the reader" and also "enhance the linguistic component of the IM message" (Lewis & Fabos, 2005, p. 483). Notably, the use of the asterisk to indicate misspelling shows attention to "correctness" and convention. The researchers demonstrated that rather than just a trivial practice, IM was instead a site where youth engaged in complex practices as they worked to meet social norms for written and spoken communication online.

Jacobs (2006, 2008) also studied instant messaging in a case study of the writing practices and school-based literacy skills of a 15-year old girl she called Lisa, and found the importance of context was heightened by use of IM:

> She was able to select words, make stylistic choices, and attend to grammar, spelling, and mechanics according to the needs of her audience, her purposes, and the requirements of genre and mode of communication. IM was one of a repertoire of literacy practices Lisa used depending on her needs. (Jacobs, 2008, p. 208)

Jacobs' work with Lisa takes into account the aesthetic choices she makes in digital composing, but also considers how digital texts are social spaces. She also recognizes how they are situated in the larger scope of her participant's literacy life, part of a "repertoire of literacy practices" she deployed depending on her needs.

Even with this new focus on social uses and social context, many researchers persisted in approaching the study of multimodal digital composing as a practice that is divorced from social context and social uses. This is not surprising, since so much of traditional literacy education has focused on end products removed from authentic contexts for composing. The process, writer's intent, or design were thus available for assessing in schooled literacy contexts, and this has been taken up by literacy educators and researchers looking to bring multimodal digital composing into schooled contexts for literacy. Leander and Boldt (2013) criticized this approach to multimodal composing as design as only accounting for a very limited part of youth's experiences with literacy, and as privileging those digital literacy practices that are utilitarian, controlled, and planned. New contexts for composing made possible by mobile technologies, for example, problematize social and cultural constructions of concepts long considered fundamental to writing and literacies, most notably, the writing process and its concomitant concepts the draft and the "final" draft.

Not only are the social elements and uses of digital multimodal texts important to consider, but for those youth composing digitally and multimodally by way of their mobile phones, mobility becomes a central affordance significant to understanding these literacy practices. The focal participants whose literacy practices are discussed in this book all used mobile phones to access the Internet and to compose digitally and multimodally. Their phones were their constant companions as they traveled across the many spaces they inhabited. This mobility, along with the particular affordances of the mobile phone hardware and the architecture of the applications they used to create and share texts, engendered literacy practices that challenge a view of composing as planned, designed, and product-focused. Instead, their mobile-phone based composing practices represent "ongoing discursive activity" (Vasudevan & Hill, 2008, p. 231).

For example, Eli told me that he only engaged in multimodal digital composing when he was bored. He often posted from the backseat of his mother's car as she drove him to and from school and after school activities. When Eli sensed he was in a state of boredom or waiting or the *in media* space of transit, he could move into a new space wherein he could chat with friends, express himself, or locate a source of entertainment. His location in space was thus related to his digital literacy practice in that one motivated the other. Many times, his literacy practice was spontaneous and reactive to the conditions of the physical space, rather than planned or focused on an end product (cf. Leander & Boldt, 2013). He would often post photographs that captured the physical environment and his musings in the moment, as seen in Figure 5.2. He engaged in an embodied literacy practice that captured the movement of his body through and across spaces.

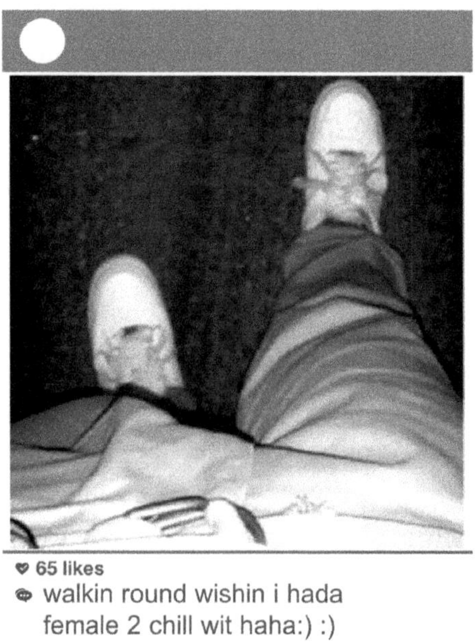

Figure 5.2: Eli's Mobile Literacy Practice.
Source: Author.

Eli's literacy practices indexed markers of his identity, conversations with friends, literacy practices such as the selfie, physical spaces themselves, and popular culture across the spaces he inhabited physically and digitally in overlapping and murky combinations. He inhabited and moved through "multiple, simultaneous space-time contexts … coordinated and produced through activity" (Leander & McKim, 2003, p. 238). His practices were fluid and not fragmented. They were part of his social practice, rather than isolated literacy accomplishments. Eli often posted pictures to his social media spaces he had taken while at school. He engaged in school-like practices outside of school such as composing narratives through photography. In school, he explored the possibilities for life and he did the same through his digital literacy practices (Moje, 2000). In and out of school, he used his phone to call upon the Google search engine as a means of locating the information he needed to accomplish moment-to-moment goals. While there was no observable planning and an absence of design as we might traditionally conceive of it, Eli's texts were at once communicative, were spaces for social engagement, and part of his larger social practices.

Temporal Scales and Design

Traditionally, the composing process and the end product have been seen as happening on a particular temporal scale. That is to say, we might think that a composer conceives of an idea for a composition, takes it through a draft and several revisions, and then the end product is submitted or published, where it sits, finished. But with the emergence of digital tools for composing, this process of planning, revising, editing, publishing has been reconfigured. This is in part a function of the affordances of new tools for composing as well as new rhetorical situations for digital composers that allow for feedback (sometimes even near-immediate feedback).

Digital media have made it easy to revise texts; whereas in the past, a finished product might be painstakingly typed onto physical paper, painted on a canvas, or exposed on expensive film that took time and money to process, with digital tools, the revision process is not only easier, requiring fewer resources, but it is so easy it can in fact be ongoing. Nothing ever has to be "finished," and in fact, with digital content, we never know for sure who our audiences are and who has possession of our texts to do with them what they wish.

Indeed, for the youth involved in this research, this was the case. A photo posted online might be "published," but if the audience reception was such that they rethought the posting of their composition, they might delete it from their digital writing environments. I noted this very early in the research, thankfully, as the disappearing data was something I had to account for in my methodology!

One day, I was carrying out participant observation on Instagram when Eli posted a photo of himself that garnered expressions of shock from his friends. After receiving feedback from several friends indicating that they were unsettled that he would post the picture, he deleted it (presumably in response to the comments). In this way, revision often came after publication in Eli's practice. In fact, he often used feedback in the space to make decisions about revising his text. He considered what others thought (addressivity) as he formed his utterance. For Bakhtin, a text does not constitute an end-point, but instead a link in the larger chain of communication, a part of this "ongoing discursive activity" (Vasudevan & Hill, 2008, p. 231). The malleable digital spaces in which Eli composed with his phone allowed for this kind of tinkering and as such it "re[wrote] the conceptual binary of process and product" (Alberti, 2008, p. 1). As DeVoss, Eidman-Aadahl, Hicks, National Writing Project (U.S.) (2010) muse, "In new digital writing environments, what is a draft exactly? Is a draft when the computer saves our work automatically? When we choose to share a version with colleagues? Or is a 'draft' something else?" (p. 57). The temporal scales on which mobile phone based composing happens certainly challenges traditional notions of design. Indeed, ongoing practice turns this notion of design on its head.

The participants' composing practices were not only ongoing, but they were also unplanned and highly reactive. The mobility of the cellphone as a tool for composing means that composing can happen anywhere. For the participants in this study, the smartphone was a constant companion, and they could snap and upload a photo to their social media accounts in seconds in a fluid, frictionless process. In this way, many times their literacy practices were spontaneous and in the moment, rather than planned or focused on an end product (cf. Leander & Boldt, 2013). Their composing practices were in dialogue with the spaces, both digital and physical, which they inhabited, as they used their phone cameras to capture physical spaces to create texts, or relied on existing voices and content in digital spaces to compose. Their practices were thus responsive and emergent rather than controlled and planned (cf. Leander & Boldt, 2013). Instead of creating planned texts from the ground up, the participants responded to the world around them, digital and physical, through texts and often revised their compositions *after* publishing, a practice that appeared to be related to the rapidity of the composition process.

For example, when I asked Eli about when or why one might post a selfie, Eli told me, "It just depends on what you feelin. When you are feelin some type of way." "Some type of way" was an expression the youth in the study used to describe a state of feeling emotional. Eli posted a picture of a framed photograph of his family, pre-divorce, himself featured as a young child. He added the caption to his digital posting, "This picture got me feelin some type of way." Since Eli relied heavily on photography as a means of composing digitally in the online space, the available means of expression were often limited to physical objects that could be photographed (in the previous example, this was the framed photograph). Oftentimes, as in his selfie practice, he used his body, photographed and posted to his Twitter or Instagram account, as a means of communicating. In this way, Eli used objects in the physical space, captured with the camera on his phone, to convey a message quickly and without having to use linguistic text to do so, perhaps because typing long messages comprised of alphabetic text on the phone is burdensome. Selfies feature prominently the face, a powerful visual conveyer of human emotional states. Eli could quickly and effectively convey "feelin some type of way" with a self-taken self-portrait posted to Twitter or Instagram. His composing practice was thus in dialogue with his emotional, social, and material world.

Indeed, reactive composing practices are increasingly more common with mobile phone-based composers. A whole genre known as the "reaction GIF" has emerged over the past decade. A reaction GIF is an animated image portraying an emotional response that can be used to reply to an existing text. They are often used in discussion threads, in text message streams, email threads, or commenting boxes on the web, and they are easy to use as a means of communicating with mobile phones because they are semiotically rich but quickly and easily accessible via apps or other mobile phone software. Some of the most common reaction GIFs are associated with the reactions "LOL," "eyeroll," "face palm," "happy," "high five," "kiss," "shrug," "yes," "sad," "wink," "thumbs up," and "no." Sites like Giphy provide whole libraries of reaction GIFs (and other GIFs representing broader ideas) categorized and tagged with descriptive hashtags like #dancing, #incredulous, #slowclap, #yawn, and even #micdrop. This highly popular practice is indicative of the reactive nature of much mobile phone-based composing.

While the users' intentions are certainly central to the determination of what temporal scales literacy practices occur on, the affordances of the mobile phone also shape composing practice when it comes to temporal scales. In walking me through his typical approach to composing digital texts, Eli described that he had an application on his phone, and this smartphone software allowed him to connect to his social networking sites without having to stop to enter his log in credentials. Nor did Eli have to take photos from outside of the application; the application could access his camera seamlessly. The affordances of the smartphone make it possible to compose rapidly and spontaneously without much forethought. There was nothing slowing him down and encouraging him to consider a composition in the smartphone application interface. The smartphone was always with him, and he could snap and upload a photo to his social media accounts in seconds in a fluid, easy process.

As was true in Eli's case, the majority of John's original (i.e., not retweeted) tweets were reactive tweets related to his emotional state in any given moment:

> Might be the happiest day if my life
> On my way to school … swear I hate ths sht
> whats the point
> there's no place like home
> you make me feel like I can fly
> feels good to be bakk home
> today oomf [one of my followers] piss me off widd da sht she was doin man
> school been funny af today
> just uh beautiful day

His postings were reactive in this way rather than planned. If he felt "some type of way" he would create a post on Twitter. Aaliyah, too, responded to existing digital content in a reactive way. She curated digital texts to communicate to her social connections about her beliefs, values, emotional states, and what content she found humorous on the Web. By subscribing to and taking in posts from teen content aggregator accounts, she reflected on what she believed and disbelieved from this world of ideas, and she grabbed and displayed that which she agreed with or liked. She was in dialogue with the texts she consumed in this way. Her internally persuasive voice built upon the a priori authoritative discourse in the spaces as she was dialogically engaged in the online space.

In fact, the practice of retweeting, which is so common on the Web, is highly reactive and indicative of this lack of planning or design. Eli's retweeting practice was, like Aaliyah's, responsive and reactive rather than planned. In addition to creating original postings on Instagram and Twitter, Eli often retweeted the tweets of others in his sphere on Twitter. Retweeting is a way of recirculating existing content. Eli explained that he saw the practice of retweeting as a way to communicate ideas though others' texts:

Researcher:	What does it mean when you retweet something on Twitter?
Eli:	Like, you just say what the person said before.
Researcher:	What's the point of it? When would you do it?
Eli:	Like, If you like a song or something, or if you like what they said, like, if somebody says somethin like, real, if they say something funny or something, real, like, catchy, then you retweet it, or something that they believe in and you believe in at the same time, then you retweet that.
Researcher:	And what's the point of doing it?
Eli:	Just to tell people, like, you and them, like, I guess like, that you like what they said, and like, you think what they said is true or whatever.

The retweeting practice, then, allowed Eli to speak through the voices of others heteroglossically (Bakhtin, 1982) with just the press of a button. He did not have to type out an entire message to express himself; instead, if he read a tweet that resonated with him, he could press the retweet button, and the tweet would post to his Twitter page quickly and fluidly. At the same time, it would be disseminated to his followers, displaying attribution to the original author as well as to Eli. It was a way to communicate the ways in which he was in dialogue with the world of ideas in which he interacted. From Eli's perspective, it was a way of showing agreement with the idea ("you think what they said is true" and "something that they believe in and you believe in at the same time"), but also affirming the speaker ("you like what they said"). His composing, then, constituted a reaction to existing ideas and voices.

In a similar vein, audience response was central to the youth's composing practice. The youth's composing practices were distributed and less cohesive than I had imagined prior undertaking this research. Bakhtin's (1982) dialogic model holds that reception and production are reciprocal and mutually constitutive in that every utterance responds to/anticipates a response. Along similar lines, the participants composing practices were informed by the engagement in digital spaces, rather than by a priori notions about what they wanted a particular text to look like or though a planning process.

Video Sharing and Design

Although none of the participants in this study used Snapchat and at the time of data collection, Instagram Stories had not yet been introduced, video sharing, so common now, had begun to show up in the participants' practices. They would upload short videos to their social media accounts, showing clips of their daily lives. Today, Snapchat video and Instagram stories are two common mobile phone-based composing practices that challenge design theory and commonly held ideas about timescales and design theory. Snapchat video and Instagram stories combine photo and video into a single reel or story available to ones followers by way of either of the two apps. While some design elements can be applied (filters, lenses, labels, and touch-screen doodles are available to enhance photos and videos), for the most part, these platforms encourage comparably raw media to be shared.

Video sharing through Snapchat, Intagram stories, and streaming video application like Periscope encourage the sharing of everyday moments, both mundane and interesting. With the advent of high-speed mobile Internet and the support of these mobile phone applications and phone hardware including high resolution cameras, the creation, distribution, and consumption of media has become nearly instantaneous. Such video, even though it is relatively raw and unprocessed is highly engaging to viewers because it offers a representation with fidelity of the subject's body language, vocal tone, perspective, and other cues in the environment. Later this year, Snapchat has planned to introduce a product called Spectacle, which is a pair of glasses that will record video from the user's perspective in 10 second bursts and will wirelessly upload to a user's Snapchat account. This trend toward point-of-view and raw video sharing is representative of the move away from design-based approaches to digital composing.

Conclusion

The kind of spontaneity that is the hallmark of mobile phone based digital composing calls into question not only how we conceive of digital composing as a process as design, but also many of our beliefs about what we value in literate practice. As Jacobs (2013) wondered,

> One question to be asked [about multimodal composing processes] is whether intentionality is a necessary aspect of design or whether design includes the spontaneous,

> random, and unexpected. Perhaps insights from the world of art, in which creativity is a mix of the intended and the serendipitous, may provide an answer. (p. 270)

Perhaps digital composing is indeed design but a new and different sort of design. Could it be that design is privileged because of the assumed importance of intentionality in composition? To that end, literacy researchers and educators may want to begin to disentangle design and temporal scales.

For example, we may need to rethink our notions of how design and timescales are entangled. Baron (2008) warned of electronic communication's "Faustian bargain of sacrificing thoughtfulness for immediacy" (pp. 198–199) because "the proliferation of writing, often done in a hurry, may be driving out the opportunity and motivation for creating carefully honed text" (p. 198). This point of view frames thoughtfulness and immediacy as mutually exclusive. Considering the data from this study and larger trends in digital composing practice, instead of a "sacrifice" of any kind, I regard the participants' and others' practices as representing a rearrangement of process. In the participants' practice, for example, they created their postings with some immediacy to provoke a response that was then considered, and the text was never "finished" but could be deleted given a particular audience response. Thoughtfulness was not gone, but instead came later than one might expect.

So what does this mean for literacy researchers and literacy educators? In my work with teachers, I have found that the salience of this question of design centers upon assessment of multimodal composing. In one camp, you have teachers who feel that the digital composition, the product, should be assessed. How well does it communicate? How effective are the multimodal choices at meaning? In another camp, you have those who feel that the process is what should be assessed. (Often, reflections written by students about their process are used as a proxy for the process). Both schools of thought call into question what should be valued when teachers bring multimodal composing practice into the schooled literacy space. For example, is a flashy, sexy digital product created by a student who has a good grasp of how to use technological tools but that is perhaps banal or uninspired in terms of content "worth" more than a digital product that is based on strong thinking and an interesting concept yet poorly executed? Should students be rewarded for technically poor products if they had a good idea but could not execute it? Ultimately, I believe this tension has never been resolved in part because of the inability of existing frameworks for assessing "writing" to map clearly on to student work. That is to say, adolescent practice is often much richer than rubrics based on state-sanctioned curriculum are able to account for.

Youth need strategies for navigating a wide array of composing environments, not just those for which the traditional writing process (pre-writing, organizing, drafting, revision) seem applicable. While traditional teaching practice has been successful at accounting for substance and technical skill, it has been less successful at considering the ways that youth digital practices go beyond what a design-based approach to multimodal composing practice can accommodate. As I saw in the practices of the three participants in this study, context matters. What does a particular composition mean in the world? What inspired its creation? How does it function within its particular communicative ecology? The digital artifacts created never stand alone (or shouldn't stand alone). It is part of a network, a social sphere. In addition, ongoing learning should be assessed. Audience reception should be accounted for but so should the students' thinking about that audience reception. What is learned that can inform the next composition?

Youth are composing in transactional environments and increasingly schools are able to afford students more authentic contexts for composing using digital writing environments. Whereas before the emphasis was placed on the artifact as a "finished product" or "final draft," malleable, audience-focused, and contextualized digital media engender a new orientation to 21st century literacy that includes a text's life in situ. In other words, digital texts have particular qualities that are closely tied to their networked qualities: they can be shared, they can be used to solicit feedback, and they engender learning well beyond their publication. It is time that we regard composing as more than the end product of a process happening on a particular ordered, controlled temporal scale, as the evidence that some cognitive process has occurred or that some worthy writerly intention is present, and begin to privilege the social uses and social nature of mobile phone based digital composing.

References

Alberti, J. (2008). The game of reading and writing: How video games reframe our understanding of literacy. *Computers & Composition*, *25*(3), 258–269.

Alvermann, D. E., & Hagood, M. C. (2000). Critical media literacy: Research, theory, and practice in "new times." *Journal of Educational Research*, 93, 193–205. doi:10.1080/00220670009598707

Atlas, M. (1979). *Addressing an audience: A study of expert–novice differences in writing* (Unpublished doctoral dissertation). Carnegie-Mellon University, Pittsburgh, PA.

Bakhtin, M. (1982). *The dialogic imagination.* Austin, TX: University of Texas Press.
Baron, N. (2008). *Always on: Language in an online and mobile world.* New York, NY: Oxford University Press.
Barton, D. (1994). *Literacy: An introduction to the ecology of written language.* Oxford: Blackwell.
Beichner, R. (1994). Multimedia editing to promote science learning. *Journal of Educational Multimedia and Hypermedia, 3*, 55–70.
Berkenkotter, C. (1981). Understanding a writer's awareness of audience. *College Composition and Communication, 32*, 388–399. doi:10.2307/356601
Cohen, M., & Riel, M. (1989). The effect of distant audiences on students' writing. *American Educational Research Journal, 26*, 143–159. doi:10.3102/00028312026002143
DeVoss, D. N., Eidman-Aadahl, E., Hicks, T., & National Writing Project (U.S.). (2010). *Because digital writing matters: Improving student writing in online and multimedia environments.* San Francisco, CA: Jossey-Bass.
Fairclough, N. (1989). *Language and power.* London: Longman.
Fairclough, N. (1992). *Critical language awareness.* London: Longman.
Flower, L. S., & Hayes, J. R. (1980a). The cognition of discovery: Defining a rhetorical problem. *College Composition and Communication, 31*, 21–33. doi:10.2307/356630
Flower, L. S., & Hayes, J. R. (1980b). The dynamics of composing: Making plans and juggling constraints. In L. Gregg & E. Steinberg (Eds.), *Cognitive processes in writing: An interdisciplinary approach* (pp. 31–50). Hillsdale, NJ: Lawrence Erlbaum.
Fountaine, S. I. (1988). Using what they know: 9-, 13-, and 18-year-olds writing for different audiences. In B. Rafth & D. Rubin (Eds.), *The social construction of written communication* (pp. 99–116). Norwood, NJ: Ablex.
Garthwait, A. (2007). Middle school hypermedia composition: A qualitative case study. *Journal of Educational Multimedia and Hypermedia, 16*, 357–375.
Gee, J. P. (1996). *Social linguistics and literacies: Ideology in discourse* (2nd ed.). New York, NY: Falmer Press.
Gee, J. P. (2000a). Discourse and sociocultural studies in reading. *Reading Online, 4*(3). Retrieved from http://www.readingonline.org
Gee, J. P. (2000b). Teenagers in new times: A new literacy studies perspective. *Journal of Adolescent & Adult Literacy, 43*, 412–420.
Heath, S. B. (1983). *Ways with words: Language, life, and work in communities and classrooms.* Cambridge, MA: Cambridge University Press.
Jacobs, G. (2006). Fast times and digital literacy: Participation roles and portfolio construction within instant messaging. *Journal of Literacy Research, 38*, 171–196.
Jacobs, G. (2008). We learn what we do: Developing a repertoire of writing practices in an instant messaging world. *Journal of Adolescent & Adult Literacy, 52*, 203–211.
Jacobs, G. E. (2013). Reimagining multiliteracies: A response to Leander and Boldt. *Journal of Adolescent & Adult Literacy, 57*, 270–273.
Karchmer-Klein, R. (2007). Audience awareness and Internet publishing: A qualitative analysis of factors influencing how fourth graders write electronic text. *Action in Teacher Education, 29*(2), 39–50. doi:10.1080/01626620.2007.10463447

Kontovourki, S. (2012). Reading leveled books in assessment-saturated classrooms: A close examination of unmarked processes of assessment. *Reading Research Quarterly*, *47*, 153–171. doi:10.1002/RRQ.014

Lankshear, C., Gee, J., Knobel, M., & Searle, C. (1997). *Changing literacies*. Philadelphia, PA: Open University Press.

Leander, K., & Boldt, G. (2013). Rereading "A pedagogy of multiliteracies": Bodies, texts, and emergence. *Journal of Literacy Research*, *45*, 22–46. doi:10.1177/1086296X12468587

Leander, K., & McKim, K. K. (2003). Tracing the everyday "sitings" of adolescents on the Internet: A strategic adaptation of ethnography across online and offline spaces. *Education, Communication & Information*, *3*, 211–240. doi:10.1080/14636310303140

Lewis, C., & Fabos, B. (2005). Instant messaging, literacies, and social identities. *Reading Research Quarterly*, *40*, 470–501. doi:10.1598/RRQ.40.4.5

Luke, A. (1998). Getting over method: Literacy teaching as work in "new times." *Language Arts*, *75*, 305–313.

Moje, E. (2000). To be part of the story: The literacy practices of gangsta adolescents. *Teachers College Record*, *102*, 652–690.

Petraglia, J. (1995). Spinning like a kite: A closer look at the pseudotransactional function of writing. *Journal of Advanced Composition*, *15*, 19–33.

Rafoth, B. A. (1985). Audience adaptation in the essays of proficient and non-proficient freshman writers. *Research in the Teaching of English*, *19*, 237–253.

Reinking, D. (1998). Synthesizing technological transformations of literacy in a post-typographic world. In D. Reinking, M. C. McKenna, L. D. Labbo, & R. D. Kieffer (Eds.), *Handbook of literacy and technology: Transformations in a post-typographic world* (pp. xi–xxx). Mahwah, NJ: Erlbaum.

Roen, D. H., & Willey, R. J. (1988). The effects of audience awareness on drafting and revising. *Research in the Teaching of English*, *22*, 75–88.

Rubin, D. L., & Rafoth, B. A. (1986). Social cognitive ability as a predictor of the quality of the expository and persuasive writing among college freshmen. *Research in the Teaching of English*, *20*, 9–21.

Scribner, S., & Cole, M. (1981). *The psychology of literacy*. Cambridge, MA: Harvard University Press.

Siegel, M. (2006). Rereading the signs: Multimodal transformations in the field of literacy education. *Language Arts*, *84*, 65–77.

Street, B. (1984). *Literacy in theory and practice*. Cambridge: Cambridge University Press.

Street, B. (1995). *Social literacies: Critical approaches to literacy in development, ethnography, and education*. London: Longman.

Street, B. (1999). The meanings of literacy. In D. Wagner, R. Venezky, & B. Street (Eds.), *Literacy: An international handbook* (pp. 34–40). Boulder, CO: Westview Press.

Street, B. (2003). What's "new" in the new literacy studies? Critical approaches to literacy in theory and practice. *Current Issues in Comparative Education*, *5*, 77–91.

Vasudevan, L., & Hill, M. L. (2008). Moving beyond dichotomies of media engagement in education. In M. L. Hill & L. Vasudevan (Eds.), *Media, learning, and sites of possibility*. New York, NY: Peter Lang Publishing.

· 6 ·

MOBILE PHONES AND VISUAL COMMUNICATION

Abstract
The ability to share photographs and other visual media, send small pictograms (e.g., emoji), and the appearance of camera-first applications like Instagram and Snapchat (which, for adolescents, have increasingly come to eclipse the use of verbocentric applications like Facebook and Twitter), has meant that mobile phone-based communication has become progressively more visual and multimodal. This chapter explores some of the practices in which the participants in this study engaged as well as some of the larger cultural practices encouraged by mobile phones wherein visual modes of communication are front and center such as emoji, selfies, GIFs, and lifestreaming on Instagram and Snapchat.

Keywords: Emoji, visual communication, Snapchat, Instagram, selfies, multimodality, lifestreaming, GIFs

The ability to share photographs and other visual media, send small pictograms such as emoji and the appearance of camera-first applications like Instagram and Snapchat (which, for adolescents, have increasingly come to eclipse the use of verbocentric applications like Facebook and Twitter), has meant that mobile phone-based communication has become progressively more visual. In this chapter, I explore some of the practices in which the participants in this study engaged as well as some of the larger cultural practices encouraged by mobile phones wherein visual modes of communication are front and center.

Using the Web and contributing to Web content requires digital reading, writing and multimodal literacy skills, or proficiency with a variety of communicative modes including linguistic, visual, kineikonic (video), gestural, spatial, and auditory (New London Group, 1996). Computer-mediated communication is comprised of linguistic text, but it also relies heavily on other visual modes and allows for the integration of sounds, video, and other communicative modes that print text cannot approximate.

As a caveat, Manovich (2001) has warned about declaring ways new media communicate "new," as there is little new about so-called new media, and consideration of new media must be carried out "in relation to a number of other areas of culture" such as other arts and media, the materiality of modern computers, today's visual culture, and how a culture encodes information: "information culture" (p. 13). Manovich's approach is in contrast to the dominant one in literacy-education research, which often attempts to equate multimodal communication to a language, a perspective influenced by scholars such as Kress (2003, 2010) and van Leeuwen (2005) who liken communication on new media to a structuralist system. New media are cultural objects (Manovich, 2001) and thus must be understood not in their discrete parts, but in their social use in an ecology that includes material and cultural aspects, including the "ways in which information is presented in different cultural sites" (Manovich, 2001, p. 13), or information culture.

There are, however, some distinctions that characterize the milieu of digital discourse. Mirzoeff (1998) declared present-day culture "ocularcentric" because people increasingly interact with visual images. Another distinction of postmodern culture relates to modes of communication: people also now interact haptically with media. The term haptic comes from the Greek, meaning "to touch," and describes the nature of touch-screen technology. This touch-screen technology is what the participants in this study used in order to create their compositions using their phones, and with it comes particular affordances and constraints.

In the next section, I use data collected from the three study participants to illustrate some of the practices in which they engaged that centered upon visual modes of communication supported by their mobile phone hardware and software, including social photosharing. I will also discuss how some of their highly localized practices are located within the larger landscape of what mobile phone-based composers are doing to communicate through visual modes with mobile phones.

Social Photosharing

A few months into this study, two of the participants, Eli and John, migrated from Twitter to Instagram, a social photo-sharing site. Instagram allows users to upload cell phone photos through the Instagram app, use filters to process and edit the photos, add linguistic (and even emoji-laden) captions, and tag the photos with categorizing or descriptive hashtags. The two posted mostly selfies to their Instagram accounts.

An integral component of the selfie is the social, addressive component: the selfie is posted online, allowing others to comment on the photo and offer feedback. For example, Eli posted selfies that captured his outfit or "swag" (i.e., style or fashion sense) on any particular day, as well as selfies that told a story, like "I'm at work." Sometimes, his selfies worked to do both (see Figure 6.3).

Figure 6.3: Eli's Work Swag Selfie.
Source: Author.

On Instagram, one's "followers" can post feedback in response to a user's postings in the form of "likes" (by pressing a picture of a heart or tapping the picture twice) or by leaving a comment (by using his or her fingertips to peck out a message in a comment box under the picture). The photo from Eli's feed in Figure 6.3 was "liked" by 60 Instagram users, an expression of acknowledgement or approval. Oftentimes, Eli's selfies would generate long lists of comments from friends as well. Lenhart and Madden (2007) asserted that adolescents use social networks to garner "feedback and affirmation" as they develop their identities. Indeed, the participants used feedback to inform their participation in the space. In fact, the participants often deleted pictures that received negative or little to no feedback. In this way, their audiences took on a role as co-constructors of the participants' texts.

Taking and posting selfies is a way of narrativizing social worlds through photography. In other words, youth who post selfies often do so to share the story of their everyday lives or to report out on their moods or emotional state. This visual storytelling is quite dialogic. It happens in networked, social spaces for audiences of friends and other connections, or if their social pages are unrestricted, the whole world. In contrast to photography of the past, which was often about capturing a moment in time or a particular scene, the self-taken self-portrait can be a practice of identity formation in addition to a narrativizing practice; as youth post selfies to social networks, they garner feedback that informs their burgeoning sense of who they are (Lenhart & Madden, 2007). The selfie sharing practice thus fosters the process of becoming. In this way, the composing practice reflects a view of the self as object and self as malleable. The participants in this research seemed to understand that the self should be informed by the social reception made evident through comments and likes.

Some of the selfies that John and Eli posted, though, seemed to be more aesthetic than anything else. In other words, they were about self-expression. In that case it is, rather than alteration or formation, the recognition of what is that is sought through the posting of a selfie. Such selfie practice is more or less also an exercise in gathering audience feedback, and that feedback informed the participants' in some particular way. Wargo (2015a) asserts that selfie practice is both about identity formation and, in part, aesthetics: "Selfies and the emerging practices of everyday digital photography are rich sites for future new media research as they demonstrate the discursive connection between creative practice and individual subjectivity" (p. 3). As youth take multiple selfies, decide among them which to post to their social pages, use filters or other processes to alter the photographs so that they are aesthetically pleasing, and then perhaps add frames, hashtags, and/or captions, they are engaging in a process that is originative, aesthetic, as well as constitutive of identity.

Social, Multimodal Composing

The feedback that the participants in this study received they seriously considered. The participants in this study often took up the feedback they received in the composing space; most often, they used feedback in the space to make decisions about revising their texts. I first took notice of this practice when, one day, I was carrying out participant observation on Instagram and Eli posted a compromising photo of himself. I watched as friend after friend posted comments that expressed shock underneath the photo. Shortly thereafter, Eli deleted the photo. In this way, he used feedback from others as he socially constructed his text. He considered what others thought (addressivity) as he revised his utterance. In this way, Eli, new to Instagram at the time, was socially learning how to interact in the space in a way that fit with social norms or garnered positive reception.

Interestingly, in addition to learning about social norms in this composing space, Eli was also doing rich composing work. He was engaging the process of revision after the publication of the text, reconfiguring the customary writing process wherein revision happens prior to publication. In this way, his audience was a co-author who had a real say in the composing process. This dialogic composing process results in a text that, though held together by the central voice of the composer, contains the voices of others, both through the comments and other feedback that permeate the space, but through the echoes of posts deleted in response to an unfavorable sort of reception.

Celebrity Selfie Culture, Corporate Interest, and Instafame

The participants were influenced not just by the overt feedback they received in the space but also by the wider culture of selfie taking and sharing online. They followed celebrities there who posted a great deal of selfies such as athlete LeBron James, singer Beyoncé, and their favorite rappers, Meek Mill and Drake. In order to better understand their wider spheres of participation, I followed those celebrities who the participants followed and over time came to recognize the ways in which the participants' postings mimicked the postings of their favorite celebrities.

For example, I was initially struck by the participants' focus on branding in their postings. They posted photos of their designer shoes, logos on backpacks, and other purchases like headphones. I saw a direct link between these kinds of posts and similar celebrity posts. Some companies leverage celebrities' large followings on Instagram to market their products using advertising methods that feel less intrusive and more organic than marketing by way of broadcast media. For example, singer Beyoncé has integrated Pepsi into her postings and athlete LeBron James endorsed brands through social media postings as well, including Nike shoes and Audemars Piguet watches. (In fact, James commands more money per post than any other athlete). John posted several images that were almost identical in content and framing to posts by LeBron James.

As is evident from these examples, the youth in this study were informed by the others voices in the spaces in which they composed. They may not have been aware of the ways in which they were influenced by the words, phrases, and visual images that came before them, but it was clear that the kinds of visual communication that they created refracted other words, phrases and visuals and that their work was, in many way, in response to the communication existing in their Web spheres. And, as Bakhtin purports, their postings anticipate future responses, as well.

Instafamous

While it is in the best commercial interest for celebrities like Beyoncé or LeBron James (and especially those with a millennial fanbase) to maintain a presence on Instagram, there are some celebrities who are especially avid users of the platform. Reality television star Kim Kardashian, for example, is described on her Wikipedia page as a "social media personality" and has the highest number of followers on Instagram at the time of this writing. She even released a book of selfies in 2015 called *Selfish*, elevating in some ways the self-taken cellphone camera portrait from low to high culture (the book was published by Rizzoli, an Italian purveyor of art books). Other key selfie practitioners such as singers Miley Cyrus and Rihanna post content similar to that of Kardashian: highly curated photographs showing the highlights (and never lowlights) of their lives. They lie on beds in expensive hotels, sit front row at fashion shows, and lounge on private yachts in stylized images taken for public consumption via their Instagram accounts.

Celebrity culture on Instagram has given rise to a phenomenon called "Instafamous" (Marwick, 2015) wherein one's fame is initiated on Instagram itself through the posting of content that resonates with the audiences there and, in many cases, leveraging connections with other users with large fanbases. The 2014 mini-documentary *Instafamous* shows the stark contrast between the middle-class teenage life of fifteen-year-old Shawn Megira who has 81,000 followers on Instagram, and his carefully curated online presence that paints a highly stylized picture of his life.

While not "instafamous," the participants in this study, too, mimicked celebrity and tabloid culture in their social photosharing practice. For example, John explained that he saw digital media as a way to amplify the reach of his fashion sense and to advertise the fact that he owned designer clothing. In this way, he used the clothing as a semiotic resource (F. Davis, 1994).

Julie: Uh, why do so many people post pictures of their shoes or their clothes?
John: Cause they think they dress nice. They want to show people that they dress nice if they got expensive shoes and stuff. Wanna brag on your swag.

"Swag," as John used it, was short for "swagger" or style. Youth have been bragging on their swag for years (think the so-called Sneakerheads who spend hundreds of dollars on the latest Air Jordan shoes to wear to school or teenage girls carrying designer handbags to class) but social media afforded John an additional outlet for showcasing his style sense.

John was also an avid online clothing shopper. He and the other participants cited the use of digital media to search for material possessions that they coveted or planned to purchase in the future:

Julie: What other sites do you visit most often?
John: Clothing websites.
Eli: Oh, yeah Amazon. I was on eBay yesterday.
Julie: What do you do on Amazon and eBay?
Aaliyah: Yeah.
John: Lookin' at shoes and stuff.
Eli: Go on Google Images.
Julie: What would you do on Google Images?
Eli: Like uh different colors of the shoes and stuff. Like when I type in Jordan there's a lot of different colors and stuff will kind of pop up.
Julie: And just to look at them?
Eli: Yeah, or you can, plan the future, if I see any I will buy.

Like the celebrities they saw posting on Instagram, the participants, and especially John, were interested in reflecting a particular lifestyle and image and keeping up with the latest trends in clothing and footwear (not unusual for adolescents, but certainly an undeniable outgrowth of Instagram and selfie culture).

Interestingly, the participants described other multimodal cues they used to project an enviable lifestyle and image. Some, like emoticons, surprised me. Traditionally, emoticons, or facial expressions created by using text, have been used to add an emotional quality to electronic communication. John often used emoij, or advanced emoticons that show not only detailed facial expressions but also can include any number of tiny pictures of objects like food, transportation, and more. In fact, out of all of the three participants and even those with whom he interacted regularly in his Web sphere, John used the most emoji. To my surprise, John explained to me that he used emoji not just to relate an emotion, but also as a way to celebrate personal assets. When I asked about the use of emoji in tweets and on Instagram, John told me, "If you have an emoji then you let them know that. You let them know that that you have a iPhone cause emojis only work with iPhones." To John, then, the use of emoji signaled not just an emotional state, but also that a person owned a particular kind of expensive and trendy smartphone. The use of the emoji was a "flex" as he called it, or a way to show off something desirable that you had.

Many of the social media accounts the youth in this study followed were those of so-called "influencers," usually high-school or college-aged tech-savvy youth who amass large followings on social media and leverage those accounts to create partnerships with brands. These youth post advertisements for brands that feel organic because they are not posting copy created by marketing teams verbatim but are creating or reformulating the content on their own in their own voices. For this reason, for many youth, the line between paid advertising and authentic testimony blurs. And weary of being advertised to by brands in ways that feel inauthentic and "corny" as the participants related, brands use this particular avenue of advertising to garner youth buy-in. While John asserted that he posted photos of his shoes, backpack, headphones, and other brand-name products because he was "bragging on [his] swag," then, it is possible that John and the other youth in this study posted brand-centric content because it is the norm in their Web spheres. It is the type of content they consumed and somewhere along the line understood as the norm for communication in that space.

Agency and Social Photosharing

This is not to say that the participants in this study mimicked, drone-like, the practices they observed in the digital spaces where they interacted. While the youth in this study did take up many of the practices common to the spaces where they interacted, they also created their own unique practices and broke away from the larger platform-based as well as social conventions. For example, although many of the teens in John's Web sphere posted copious selfies, what was unique to John's practice was his affinity for illustrating song lyrics through the photos he posted. John would take a selfie, then match the selfie to a hip hop song lyric, as demonstrated in Figure 6.4.

Figure 6.4: An Illustrated Song Lyric Plus Selfie.
Source: Author.

John explained that when a particular song lyric resonated with him (when he "feel[s] it") he would post the lyric, either with a selfie previously taken and already in the phone, or by taking one on the spot. John's illustrated lyric postings are an interesting exercise in transmediation (Siegel, 1984) as he represented an aural artifact visually and through linguistic text, extracting salient lines from the lyrics of his focal songs. He used the song lyrics and his photography practice to express his emotions and his identity, and to give his network information about his way of being in the world (cf. Freire, 1972; Street, 1984). This was a heteroglossic practice wherein two social languages populated the same utterance (Bakhtin, 1982). In addition, he used linguistic text to evoke aural information, demonstrating an understanding of the ways in which the modes were in dialogue. For example, in Figure 6.4, which features two selfies and lyrics from rapper Drake's song "Girls Love Beyoncé," he uses the hashtag #SayMyName, the name of the song from which these lyrics are mined. In this way, he labeled the post as a reference to the song and made his post searchable to all users of Instagram who enter this hashtag in the search function.

Eli also created his own visual communicative conventions. He interchanged numbers with letters, often to such a degree that his composing was practically indecipherable to me. He almost always used the numeral "6" in place of the letter b. When I asked him about this, he said he simply replaced his bs with 6s because it looked cool.

Julie:	Uh, what about when people use letters wrong? Like I kept seeing um like a q for a g. Like you would …
John:	Or a 6 for a b.
Julie:	Yes, what's the point of that?
John:	That's jus' how people design it.
Eli:	Oh, I'll do that.
John:	Some people jus' design their own writin' like that.
Eli:	I'll do that with my own my own Twitter name [obscured]. I replaced the um e with a 3.
Julie:	Why?
Eli:	It's just funny to read, man. Cause it looks cool.
John:	The 3 looks like a e, the 6 looks like a g …
Eli:	The 6 looks like a b.
John:	The 6 looks like a b. And the 9 look like a g. So they jus …
Julie:	They're just being creative?
John:	Yeah, that's exactly what it is, being creative.

Among his connections, Eli had a distinct style; I followed his social connections on Twitter, and over the course of the study through participant observation I did not see anyone else using this convention. This visual mode seemed to be for him a means of self-expression and for ongoing identity performance, as it was for John.

Visual Communication with Smartphone Cameras

As the selfie practice indicates, the participants relied heavily on the cameras on their phones to compose. That meant that they relied on objects in their physical worlds to communicate. In this example from John's Twitter profile, he engaged in wordplay using physical objects (see Figure 6.5). John arranged a loaf of bread and a stack of money on his kitchen counter and then took a photograph, playing on the slang term for money, "bread." In this way, he could use physical objects to communicate (in this case, since he used the photograph as his profile picture, it seems to be intended to mean that he has a lot of "bread" and that he is also linguistically clever).

Figure 6.5: John's Twitter Profile Page, Showing Wordplay with Physical Objects. Source: Author.

One's own body, too, is a convenient means of multimodal communication when the mobile phone camera is the composer's primary tool. Eli often posted selfies that showed off his body, particularly his abdomen. As an athlete, he was committed to physical fitness, and also enjoyed performing a masculine subjectivity. Selfies communicate multimodally using the body, primarily. In most of the selfies he posted, Eli donned a hat worn too large, backward facing or cocked to the side, a trendy way of wearing a baseball hat. He communicates about his sense of style. Wearing a hat impractically, with the bill to the side or back, demonstrates a commitment to style over functionality.

Although the photos seem to be at least somewhat stylized, Eli attempts to look casual and as if he has not invested too much effort in taking the photograph (often the photos were taken askew and off centered as if to suggest fast composition, for example). This phenomenon of the planned but seemingly off-the-cuff self-portrait has been called the "casual selfie." While it is more spontaneous and casual than a traditional photographic studio self-portrait, the selfie is planned and staged to some degree. Eli must use a mirror to take these photographs, and he must review and approve of the photo before posting. The participants described taking several versions of a photo before choosing which one to post to their social media pages.

Selfies are tantamount to bricolage (Lévi-Strauss, 1966) because they engage objects in the physical space to communicate. The camera as a means of multimodal composing seems to influence this approach of using what is available and convenient to be captured and displayed for communication. As was explored in Chapter 3, the physical and material worlds of youth are digitized through mobile-phone based composing practices such as cellphone photography and thus afford semiotic resources for meaning-making in the digital space.

While the participants in this study did not usually employ them, many users of Instagram use the application's built-in filters. Filters can be applied to a photo to change the look and feel by changing the sharpness, brightness, color, glow, exposure, tone, or other effects. The Instagram filter became such a cultural trend that *Time Magazine* reported that "Parents Are Naming Their Babies After Instagram Filters" and a popular one hit wonder by the group the Chainsmokers called #SELFIE featured a line in which a clubgoer asks, "Can you guys help me pick a filter?/I don't know if I should go with XX Pro or Valencia/I wanna look tan." The names of Instagram's built-in filters are recognizable, pop-culture signposts.

In reaction to the pervasive use of the Instagram filter, the hashtag "#nofilter" became a common way to indicate that one had not altered or run through any photo editing filters their Instagram posting. It is a way of indicating that something is naturally beautiful and did not require any change before posting. It can also be a way of poking fun at the vanity that underlies the use of Instagram filters.

There are also apps available to edit photos beyond what Instagram offers within the platform. For example, the Pic Collage app allows users to stitch together several photos into a collage, add borders, text, graphics or stickers, and other effects, using a pre-existing template or by creating one's own design. The app InstaBeauty is one of many apps that is intended to "beautify" one's photos through photoediting tricks like enlarging eyes, slimming the face, lifting and slimming the nose, and removing blemishes and eyebags.

In addition to the use of filters and apps, metadata can be added to Instagram postings. This is usually done through the GPS sensors that detect location and allow users to signify at what event or geographic location they appear in or feature in a particular photo. Hashtags, or keyword labels for postings that can signal the content of the photo (or be used otherwise as commentary that compliments the photo in some way) can also be used as a form of metadata. Metadata is a semiotic affordance that communicates information to audiences and is often quite central to the composing practices of youth.

While the convenience of the built-in camera within a mobile phone supports spur-of-the-moment photography, it is certainly true that a user of an application such as Instagram can employ a number of design elements in the way of filters, photo-editing apps' affordances, or metadata. While the participants in this study rarely demonstrated attention to design in this way, it is not uncommon for mobile phone visual composers to put time and effort into crafting their postings to social networks. In fact, the taking of multiple pictures to review and choose the "best" one before posting can be viewed as a process of design (albeit a process of design that relies on a bit of serendipity).

Global, Local, Lifestreaming: Selfies and Intimacy

Social photosharing capabilities support the practice of lifestreaming (Freeman, 1997). A lifestream is a time-ordered text stream that acts as a "technologization of the self" (Wargo, 2015a, p. 2), a sort of chronological diary comprised of digital artifacts, usually images and linguistic text. Lifestreaming can be accomplished by way of long form composing like blogging, microblogging on platforms like Twitter, through images on a platform such as Tumblr or Instagram, or through more ephemeral modes, like the disappearing media within an app like Snapchat. The mobile phone encourages lifestreaming because it supports amateur, everyday photography along with fluid, unfettered sharing through apps installed on the phone.

A lifestream gives an outsider a glimpse of another person's experience. The participants in this study posted photos and other media when they were alone, and they often had emotional content and were quite intimate in terms of content and their aesthetic qualities. For example, John used the elements he had for multimodal expression to create intimacy; he took most of his pictures from close range and he took and posted photographs in private areas of his home and wider life, such as family spaces and at family events. Eli posted a video on Instagram of himself and his mother going through a box of old family photos and reminiscing. While there have been many sweeping claims around how networked media has made possible a "flat" and globalized world, in many ways, selfie-sharing practices are highly localized ones.

Because cellphone cameras offer a means of composing digitally without typing long messages (cumbersome to do on a small or digital keyboard), they offer youth a powerful means of communication. At the same time, youth are limited to the available means of expression and are often also limited, when using cameras, to objects in their immediate physical worlds that can be photographed, frequently their bodies and faces. Since the face is a powerful visual conveyer of human emotional states, the selfie is a means of quickly and effectively conveying information about the self. Youth also photograph objects in the spaces they inhabit to share online, oftentimes stylizing and otherwise arranging objects to communicate intended messages. Their composing practices are thus in dialogue with their immediate material worlds rather than globalized or otherwise distributed.

This kind of highly localized practice is made evident by the highly popular so-called "travel selfie" wherein people post photos of themselves in front of monuments, historical sites, breathtaking scenery, or other sites to which they have traveled. (The selfie stick, a handheld, stick monopod for often used for taking travel selfies because they support photography from a greater range, has become associated with tourists and has even been banned in some tourist destinations like Disney World). In these selfies, the location is featured in a way that it appears secondary to the person taking the photo.

Though it was not a technology taken up by the participants in this study, Snapchat, a photo and video messaging service accessed through a cell phone application on an individual's phone, represents this trend toward localized media. Photos and videos sent through Snapchat known as "snaps" can be sent directly to friends or can be posted for all followers to view in a format known as a "story." Snapchat stories tend to feature chronologically ordered events from someone's day. Since a story expires from the site after 24 hours, the story function encourages users to publish mundane, daily-life content. The result is that users compose on Snapchat through a series of short photos and videos of their daily activities.

Wargo (2015b) points to the affective and embodied literacy practices inherent to Snapchat composing. He argues that literacy researchers, instead of attending solely to what's happening behind the screen should look at literacy practices happening *through* technology. As users compose with Snapchat, the mobile phone becomes an extension of their perceptive faculties; the audience sees what they are seeing and is, through verbal commentary or written (or emoji) annotations, also privy to their thoughts about what they are experiencing. This digital literacy practice is so localized, as Wargo (2015b) describes his participants' Snapchat composing, it requires the audience to "be a part of the performance rather than decode it from a periphery" (p. 49).

Snapchat is also localized in that the user can better control who receives or views their multimodal composition. Whereas on Twitter or Instagram, even if one "locks" or privatizes their page, it is impossible to know with any precision who the audience is for a particular posting or how many times they have viewed the posting. On Snapchat, postings can be sent directly and the composer can control for how much time the posting is made visible (up to 10 seconds). As boyd (2014) points out, this is the only platform that notifies the sender when their content has been viewed (notable in a media landscape wherein people use the like button to "like" "content that is inherently unlikeable just to acknowledge that they've consumed it).

With highly localized new applications gaining popularity (such as social network Yik Yak, dating app Tinder, and even mapping app Waze, which rely on geolocation software to allow users to communicate with those in their proximity), in the coming years we may see mobile-phone based communication becoming less public and more controlled. Whereas the dawn of the 21st century saw much discussion of the coming of a "flat" and globalized connected planet, it seems that in reality users have harnessed digital tools to share localized practices.

Emoji

Back in the 1990s when I used my beeper, described in the preface, to convey emotional content in a fast, codified way, it was through numerical codes. For example, the code 143 stood for "I love you" because it represented the number of letters in each word. Other messages one wished to send required that the beeper be flipped upside down, as the letters were visually suggested by the look of the numbers used when flipped (for example, 1134 2 09 for "go to hell"). As beepers became obsolete, emoticons (a portmanteau of emotion and icon) appeared in email and text communication, facial expressions created by using text, such as ;) :D or :-(, employed to add an emotional quality to electronic communication. Such communicative elements are necessitated by the absence of prosody, kinesics, and other visual cues afforded by face-to-face communicative formats and were comprised of American Standard Code for Information Interchange (ASCII), the most common character-encoding scheme. Apple's iPhone (and later Android) included emoji as part of a fonts package in the 2000s and emoticons lost traction in favor of these small pictures which were more visually vivid, detailed, and include a wider variety of visuals from smiley faces to flags to foods.

Perhaps the role that emoji have come to play in modern communication is most aptly demonstrated by the fact that an emoji, the "face with tears of joy" smiley, was named Oxford English Dictionary's Word of the Year in 2015. In their 2015 emoji usage report, marketer Emogi reported that 92% of people online use emoji, one-third of those users use them daily. Emoji replace linguistic text and make it possible not only to convey emotional content (*Wired Magazine* called emoji "the new language of the heart") but also to communicate information more quickly since the information is visual and thus does not need to be decoded like linguistic text. In fact, emoji can be understood across languages, enabling, to some extent, global communication. In the same way that photographs can be used to convey a lot of information more quickly than that information can be typed out using verbocentric methods, emoji work to provide a more frictionless means of conveying information on a mobile phone.

GIFs

In addition to static visuals such as photos and emoji, mobile-phone based composing has come to integrate moving visuals as well, usually in the form of Graphic Interchange Format files, or GIFs. This bitmap file format is small and uses very little data and is thus suitable for simple image exchange, and this is why it has increased in use and popularity with the rise of mobile phone-based communication. GIFs are not a new file format, however. The GIF was invented in 1987 by CompuServe's Steve Wilhite; the first one was of an animated airplane. GIFs have stuck around as a file format in use since that time, but only fairly recently have they migrated from blog sites like Tumblr and Buzzfeed to be easily accessible to mobile-based composers writing on-the-go.

Twitter was the first major social platform that integrated support for photo and video after offering support for only text-based communication initially and Facebook eventually followed suit. They later, in that same order, offered support for GIFs (Facebook initially resisted so as to not make users' newsfeeds "too chaotic"). As of this writing, the newest iPhone software update included a native GIF keyboard with a keyword-searchable GIF library offering thousands upon thousands of two to three-second instant-play animated images that can be easily integrated into text messages. GIFs can thus substitute for linguistic text or they can enhance linguistic text.

GIFs usually feature clips or animated stills from television shows, movies, celebrity footage, or other popular culture elements that are immediately understandable not just because they are visual but because they allude to recognizable cultural messages or touchstones. They are soundless, but are sometimes overlaid with text. The GIF is much easier to cognitively process than even just an image, usually because it uses that popular culture reference to add more dimension to the visual and can thus register with readers more quickly by way of that shared cultural knowledge. Compared to linguistic text, they can be a less clunky means of communicating complex emotional content like sarcasm, for example, which can be sometimes difficult to decode in written form. For communicating tone and nonverbal language, GIFs are useful for mobile phone composers.

Marketers refer to content like GIFs as "snackable" because they are bite-sized or easily and quickly consumed. Mobile phone-based communication, which is usually consumed in such bite-sized bits of time, such as when someone is waiting in line at the grocery store or on an elevator, may be driving the move to snackable communication. In 2016, for example, *Business Insider* reported that 60% of time spent on social media happens through mobile devices. The more efficient and potent communication is that is being consumed on-the-go, the more effective that communication is, particularly when it is being consumed on small screens like those of most mobile phones. It is much faster to consume a visual representation than a linguistic explanation of that same idea.

But beyond ease of reading, GIFs are useful for mobile phone-based composers looking for more precision of expression with regard to emotional content. GIFs can enable users to express emotions with more granularity than what emoji or even verbal language might offer. For example, there is a difference between in signification between "anger" and "rage" or "sadness" and "despondency" and GIFs are better suited for expressing a full range of emotions through their animation and through their use of outside cultural references. (You can test this by going to the site GIFGIF at gif.media.mit.edu and querying the tool for a GIF that's equal parts afraid, disgusted, and amused). Not to mention, with GIF keyboards and apps making it possible to access vast libraries of GIFs (and with some apps you can create your own with your original photographs) whereas the typical emoji keyboard taps into 1,620 emoji.

Mobile Phones and Postsymbolic Communication

With the particular toolkits the mobile phone has afforded, the trend has been toward postsymbolic communication. In his 2010 book, *You Are Not a Gadget*, Lanier describes such communication as that which goes beyond symbolic or codified communication, not "an annihilation of language as we know it" but "a vivid expansion of meaning" (p. 190). Although emojis and photographs are still in a sense symbolic, they are a more direct means of representation than linguistic text.

Snapchat is one phone-based platform that moves closer to the kind of communication that Lanier describes. By "cutting out the 'middleman' of symbols and directly creating shared experience" in "a fluid kind of concreteness [which] might turn out to be more expressive than abstraction" (p. 190), Snapchat (and now Instagram Stories, which offers a similar "portal" into the daily lives of youth) is the most postsymbolic of the platforms that are in widespread use by youth. Mobile phones, it may be argued, are increasingly giving youth a more seamless, direct, and perhaps even more human way to communicate across spaces that is in some ways less mediated than we might expect of an advanced technological tool.

Discussion

Means of visual communication particular to the mobile phone such as photography, emoji, and GIFs allow youth to compose in ways that go beyond the limitations of alphabetic text alone. Visual mobile phone based composing practices are in many ways less abstract than composition that relies upon codified linguistic text. For example, where language might be used to suggest and approximate emotional states, describe physical objects with words that represent those objects, or to signify ideas, a photograph, video, emoji, or other visual element can be used as a means of representation that does not rely upon codes or signification. In this way, mobile phone-based communication can represent a very direct means of communicating ideas for youth that removes the "middleman" of linguistic codes.

This is not to say that youth do not engage in practices that involve styling or otherwise revising and refining their compositions using mobile phones. In addition to using feedback in the digital space to learn more about the preferences of their audiences as a means of informing their composing process, youth use filers, metadata, and editing apps to revise, frame, or to complement in some way their visual compositions. In addition, they often use a mashup of visual text, linguistic text, and metadata to communicate ideas, emotions, or to represent their physical and material worlds.

Perhaps the hallmark of mobile phone-based visual communication is the selfie. The popularity of the selfie as a visual, mobile phone-based composing practice has given rise to the cult of social media personalities such as Kim Kardashian and the so-called "Instafamous" set and concomitant commercial influence. Because youth learn how to compose in digital spaces by observing and participating in those spaces, they sometimes take up and employ commercialized discourses like those prevalent within selfie culture.

Selfies, at once communicative and aesthetic, allow for means of communication for youth that go beyond abstract representation through the code of linguistic text. The participants in this study, for example, married the aesthetic and the communicative. In this way, mobile phone-based visual communication has given rise to a new norm for composing that marries visual appeal with other elements that are purely communicative.

References

Bakhtin, M. (1982). *The dialogic imagination*. Austin, TX: University of Texas Press.

boyd, d. (2014, March 21). Why snapchat is valuable: It's all about attention. Retrieved from http://www.zephoria.org/thoughts/archives/2014/03/21/snapchat-attention.html

Davis, F. (1994). *Fashion, culture, and identity*. Chicago, IL: University of Chicago Press.

Freeman, E. (1997). The lifestreams software architecture. Retrieved from ProQuest Digital Dissertations.

Freire, P. (1972). *Pedagogy of the oppressed*. Harmondsworth: Penguin.

Kress, G. (2003). *Literacy in the new media age*. London, UK: Routledge.

Kress, G. (2010). *Multimodality: A social semiotic approach to contemporary communication*. London, UK: Routledge.

Lanier, J. (2010). *You are not a gadget*. New York, NY: Knopf.

Lenhart, A., & Madden, M. (2007, April 18). Teens, privacy, and online social networks. Retrieved from http://www.pewinternet.org/files/old-media/Files/Reports/2007/PIP_Teens_Privacy_SNS_Report_Final.pdf.pdf

Lévi-Strauss, C. (1966). *The savage mind*. Chicago, IL: University of Chicago Press.

Manovich, L. (2001). *The language of new media*. Cambridge, MA: MIT Press.

Marwick, A. (2015). Instafame: Luxury selfies in the attention economy. *Public Culture*, *27*(1), 137–160.

Mirzoeff, N. (1998). *The visual culture reader*. London: Routledge.

New London Group. (1996). A pedagogy of multiliteracies: Designing social futures. *Harvard Educational Review*, 66, 60–92.

Siegel, M. (1984). *Reading as signification* (Unpublished doctoral dissertation). Indiana University, Bloomington, IN.

Street, B. (1984). *Literacy in theory and practice*. Cambridge: Cambridge University Press.

Wargo, J. M. (2015a). "Every selfie tells a story..." LGBTQ youth lifestreams and new media narratives as connective identity texts. *New Media and Society*, 1–19.

Wargo, J. M. (2015b). Spatial stories with nomadic narrators. *Journal of Language and Literacy Education*, *11*(1).

· 7 ·

DIGITAL CURATION

Abstract

This chapter defines the practice of digital curation and discusses the ways in which curation is a practice that is easy to engage in with the mobile phone because so much semiotic richness can be accessed with the touch of a button. It shows how the retweeting function in Twitter allows composers to speak through others' voices with the press of a button (which is much easier than typing a long-form text with the digital keyboard) and engages bricolage and remix. It explores the ways in which this communication is layered: it allows the youth to show agreement with the idea as well as to signal alignment with or to affirm the original poster. Through text curation, the practices of reading and composing are almost collapsed. There is no observed planning, and instead the process is a continuous, fluid one, one in which reader and writer collide, blurring the boundaries between author and audience. This chapter also explores highly inventive youth practices using curation tools.

Keywords: Digital curation, curation, retweet, bricolage, remix, Twitter, curation tools, mobile phone

This chapter explores the practice of text curation, colloquially known as "secondary content promotion," "reblogging," "retweeting," or "pinning." In this practice of digital curation, digital composers create or recontextualize a text using others' content quickly and fluidly and without permissions or citation. This practice of bricolage or remix is becoming increasingly commonplace around the web; in fact, sites such as Tumblr and Pinterest have been developed just for curating. Even Facebook and Twitter, the two most popular social networking sites, have features that facilitate digital curation.

As even a casual user of the web knows, the concepts of authorship and the ownership of ideas or content are very different in the realm of the digital and the world schooled literacy, which adheres for the most part to print literacy-based values and conventions. The values inherent to new digital composing practices, however, bleed into young people's composing practice in school, and many teachers lament this fact quite vocally; this generation has earned the nickname the "cut and paste generation" because so many teachers are finding that students either copy wholesale or piecemeal digital content from around the Web and then insert the material into their school assignments. While for many who are not entrenched in this digital world with its concomitant new ethos around authorship, sharing, and remix, this seems to constitute a moral or ethical violation. For teachers, this practice of curation as it exists on the web is not just unsanctioned, but breaches academic norms and rules. In fact, in the academic sphere, the concept of the appropriation of intellectual property is discussed as an issue of plagiarism. On the web, however, curation is a form of authorship.

Even after the "social turn" (Gee, 1996) which occurred in the 1980s wherein thinking around how literacy develops moved from a focus on cognition to discourse communities, the idea of literacy as an individual achievement continues to persist today in settings for schooled literacy. We see this in the way schools are organized. Measures of literacy are administered to individuals, even if literacy learning happens cooperatively through group activities, literature circles, or whole-class discussions. Probably because literacy has long been thought of as an act of the mind, a cognitive shift that occurs on a student level, an orientation to literacy as individual accomplishment dominates in spite of the near universality of instructional methods drawing on social theories of learning that engage collaborative approaches to teaching and learning.

In some regards, though, literacy teaching has accounted for the social nature of knowledge construction. The research paper, a popular writing genre that is part of almost every curriculum across subject areas, is one such classroom activity. Teacher preparation programs have begun to prepare teachers to employ strategies that get students out of seated rows and into collaborative circles. Peer review is widely employed in the teaching of writing process. Yet, plagiarism and attribution are still ideas that continue to endure; these are ideas that in many ways rub up against the ethos of writing online, and, especially, I argue, composing with and through mobile phones.

In terms of implications for schooled literacy learning, digital curation has significance in terms of two major ideas. The first is authorship. Digital composing practices have occasioned changes in this area because with the ability to more easily share or publish content online, content has proliferated. Digital content is so abundant it feels (and in many ways is) freely available. The ownership or authorship associated with digital content is thus tenuous. In addition, the proliferation of content has spurred the need for people to be able to organize and save the content they want to keep amid the data deluge. The second (and interrelated) idea centers on publication. Publication in digital space is much more iterative. Traditional (print) publication results in texts that are much less malleable than digital texts. The digital texts online are easier to share and/or remix the resulting curated texts are multivoiced.

This chapter begins with a theoretical exploration of the practice of digital curation, including illustrations using data from my research of youth's mobile phone-based composing practices and ends with a discussion of what the practice of digital curation means for teaching literacy in schools given that youth are, by way of their heavy interaction in digital spaces, being socialized into spaces with a very different orientation toward authorship, content ownership, and other concepts central to in-school composing.

All three participants in this study engaged the ethos of curation. Their writing practices were highly social. Twitter and Instagram handles were references to songs written by others. Media galleries associated with their profiles included pictures of places they had visited and meals they had eaten and as such they grabbed and used others' accomplishments to express their own identities. A photo someone else posted online could easily become a background image for a profile, and the participants changed elements like background images, profile pictures, and online handles frequently. Such digital curation is reminiscent of Bakhtin's (1986) ideas about the heteroglossic nature of language. Aligned with Bakhtin (1986), people select their language from other language in use. As participants selected from and appropriated others' digital texts in their Web spheres, they demonstrated the heteroglossic nature of all communication. Although the kinds of writing students were asked to engage in in school were solo-authored, online they spoke with and through others heteroglossically (Bakhtin, 1982). (Their school writing was heteroglossic as well, but this was not a feature of language sanctioned in school literacy). In school, such appropriation constitutes plagiarism or cheating, but online, appropriating the ideas and products of others is prevalent and even encouraged. In fact, on the Web, this kind of appropriation is a form of authorship. According to Jenkins et al. (2006a):

> The digital remixing of media content makes visible the degree to which all cultural expression builds on what has come before it. Appropriation is understood here as a process by which students learn by taking culture apart and putting it back together. (p. 55)

Through social writing practices like digital curation and meme-sharing, they engaged in rich literacy learning.

The remainder of this chapter illustrates the ways in which the varied digital curation practices in which the participants in this study engaged were highly social practices that relied on the voices of others, practices that are endemic online and are indeed hallmarks of digital, open culture.

Digital Curation

All three of the participants in this research engaged in digital curation. Digital curation is a practice wherein one grabs existing digital content and displays it in one's own Web space. The term digital curation is based on the word curate, which traditionally referred to physical objects and meant to collect, sort, and arrange them for display. Digital curation refers to the selection, collection, arrangement, and display of digital content online, most commonly images, links, and music or video files. This practice has developed alongside the explosion of Web content as a way for people to manage the deluge of digital information that characterizes the present moment. Many popular Web platforms (including Facebook) now feature curation tools, usually in the form of buttons that allow users to collect and display Web content fluidly and rapidly with just one click. A common practice online, A Pew Internet and American Life study reported that 56% of Internet users engage in "at least one of the creating or curating activities [the researchers] studied and 32% of Internet users do both creating and curating activities" (Brenner, 2013). On Twitter, digital curation is called "retweeting"; a user can pass on another user's tweet by pressing a retweet button. Retweeting content also copies that content to the page of the retweeting party.

Like microblogging, the practice of digital curation is popular with smartphone-using teens. With the click of a button, one can post a tweet via digital curation. Because of this lack of friction, microblogging and digital curation are easy to engage in using a smartphone. And consistent with the increase in smartphone usage, content manufacturers are designing content compatible with the ways mobile phone users are consuming and sharing content, which has both encouraged and has been influenced by youth practice.

Retweeting as Digital Curation

On Twitter, the participants all engaged in the practice of retweeting, a practice common on Twitter which is in fact built in to the interface with a button for retweeting. Eli explained that he saw the practice of retweeting as a way to communicate ideas though others' texts:

Julie: What does it mean when you retweet something on Twitter?
Eli: Like, you just say what the person said before.
Julie: What's the point of it? When would you do it?
Eli: Like, If you like a song or something, or if you like what they said, like, if somebody says somethin like, real, if they say something funny or something, real, like, catchy, then you retweet it, or something that they believe in and you believe in at the same time, then you retweet that.
Julie: And what's the point of doing it?
Eli: Just to tell people, like, you and them, like, I guess like, that you like what they said, and like, you think what they said is true or whatever.

The retweeting practice, then, allowed Eli to speak through the voices of others heteroglossically (Bakhtin, 1982) with just the press of a button. He did not have to type out an entire message to express himself; instead, if he read a tweet that resonated with him, he could press the retweet button, and the tweet would post to his Twitter page quickly and fluidly. At the same time, it would be disseminated to his followers, displaying attribution to the original author as well as to Eli. It was a way to communicate the ways he was in dialogue with the world of ideas in which he interacted. From Eli's perspective, it was a way of showing agreement with the idea ("you think what they said is true" and "something that they believe in and you believe in at the same time"), but also affirming the speaker ("you like what they said").

Eli also curated digital texts related to pop culture. Eli was interested in pop culture knowledge and liked to read online about his favorite rappers like

Meek Mill, Lil' Boosie, and Lil' Wayne. There was some drama involved in these stories that he found entertaining. He was very interested in finding out why certain rappers were in jail, and who had "beef" with who. He would conduct "research," as he put it. I inquired as to what kind of research he had done lately and he described trying to find out why Lil' Wayne had had a seizure. In conducting observations in the class and talking to the participants in the study, I came to understand that insider knowledge related to popular hip hop artists, the latest fashion including hard-to-find Air Jordan athletic shoes and other limited-edition clothing, and cultural knowledge of the latest releases via the HotNewHipHop or LiveMixTapes Twitter feeds was social currency. By subscribing to the Twitter feeds of popular artists and entertainment news outlets, looking up clothing online, staying abreast of new music, conducting "research," and recirculating content he found to be of interest, he was gaining social capital that was a means of gaining power in his world (Bourdieu & Passeron, 1990).

Likewise, John followed and recirculated content from a wide array of Twitter users, including celebrities, musicians, and Twitter accounts that were cultural aggregators, such as "Sex, Quotes, N Humor," an account that claims to post "SEX, Life quotes, and Funny Sh*t all rolled into one." He followed media companies such as ESPN, BET, and YouTube. John pointed out that he also followed Forbes, which he told me was because he believed, "If you read a bunch of nonsense every day, that's what's going to come out." He seemed to have a nuanced sense of the social, dialogic nature of composing in the digital space.

Curation through Quotation

The participants engaged in other digital curation practices outside of retweeting. For example, both Eli and John consistently tweeted music lyrics from popular songs as conventional (not retweeted) tweets. In other words, they typed lines from songs they liked into their phones, posting the lines to their respective Twitter pages. When I asked Eli about why he did this, he said one would tweet lyrics if "it's your favorite song" or if "it's the favorite part of the song." In this way, through curation was able to polyphonically express himself through the voices of others (Bakhtin, 1982).

Though she did post some original tweets, Aaliyah's Twitter feed was rife with retweeted or curated content. In fact, Aaliyah was unique among the three focal participants as a heavy circulator of content. 33% of Aaliyah's overall tweets were retweets. By comparison, 10% of John's tweets were curated, and 11% of Eli's tweets were curated. The phone Aaliyah used to connect to the Internet was Internet-enabled, but not a smartphone with the powerful capabilities of the touchscreen phones the other two participants had. It was therefore more cumbersome if not impossible to create complex multimodal compositions with the phone, but easy to curate texts, because that could be accomplished with one click. This may explain why she engaged in digital curation more often than the other two participants who had more sophisticated mobile phones.

Like Eli and John, Aaliyah subscribed to youth and pop culture Twitter aggregator accounts. These accounts took popular content from around the web to use as the basis for their feeds. Aaliyah would read tweets from these various aggregator accounts as the tweets appeared in her feed. She would then retweet the content she found relevant, true, or amusing, according to her description. Aaliyah was thus an aggregator and disseminator of content herself.

Sometimes Aaliyah would preserve the content she retweeted in its original form, but as can be seen in Figure 7.6, she would sometimes add her own commentary, thus recontextualizing the content. In Figure 7.6, Aaliyah added the commentary "LMAO" (Internet slang for laughing my ass off). In adding commentary, she shifts the previous meaning or significance of the text. It goes from being a commentary about the sartorial choices of Superman to being about her identity (i.e., what entertains her). Conversation is inherent to the practice (Bakhtin, 1986). Not only did she speak back to the original poster through a retweet with added commentary, but she also spoke to readers of her Twitter page who would get a sense of her opinions about the tweets she recirculated by reading this polyphonic curated text.

Figure 7.6: A Meme Retweeted by Aaliyah.
Source: Author.

As previously stated, Aaliyah did most of her composing on Twitter on her phone. Her phone had limited affordances for digital composing because the keyboard was small and hard to type with and because the phone had no mouse for fine manipulation. However, curation is a form of composing that is easy to engage in with even the most rudimentary mobile phone because there is so much semiotic richness that can be accessed with just the click of a (retweet) button. In Aaliyah's curation practice, the practices of reading and composing were nearly collapsed, and the process was a continuous, fluid one, one in which reader and writer collided, the boundaries between author and audience blurring.

In Figure 7.7, Aaliyah used the meme created by someone else and circulated by the account "@GhettoChildren," who added the comment "SERIOUSLY." The creator of the meme recontextualizes actor Bill Cosby's facial expression from a scene in The Cosby Show to create a message of disapproval directed toward people who wear leggings every day. Aaliyah then took up the image and used it to communicate that she finds it salient and relevant to her own perspective, and also does not approve of people who wear leggings every day. The multimodal artifact, which employs linguistic text, also uses

Bill Cosby's facial expression to communicate, almost as an ellipsis between the two parts of the linguistic text (Milner, 2013). Context also factors into the semiotic meanings of this artifact; Bill Cosby has since been the subject of a serious investigation and his likeness carries much different connotations now.

Figure 7.7: Another Meme Retweeted by Aaliyah.
Source: Author.

The data collected in this study showed that Aaliyah curated digital texts to communicate to her social connections about her beliefs, values, emotional states, and sense of humor. By subscribing to and taking in posts from teen content aggregator accounts, she reflected on what she believed and disbelieved from this world of ideas, and she grabbed and displayed that which she agreed with or liked. She was in dialogue with the texts she consumed in this way. Her internally persuasive voice built upon the a priori authoritative discourse in the spaces as she both constructed and displayed markers of her identity in a process of, to borrow from Bakhtin, "ideological becoming."

Repurposing Curation Tools

Most people use the retweeting feature on Twitter to spread content they find noteworthy to all of their followers. For this reason, I was puzzled when I noticed that John would often retweet seemingly insignificant content. For example, he would retweet others' tweets that were solely comprised of short greetings such as "hi" or "good morning." When I asked him why one might retweet (thus spreading) such a seemingly unimportant message, he told me, "[Why] retweet it? Because they feelin' the same mood as you. You know, they jus' gettin' on Twitter." Instead of retweeting to spread content—the intended function of the retweet feature—John told me he would retweet a tweet if he "feel[s]" it. In other words, if the tweet by the original poster resonated with him, he would retweet it. In doing so, he would communicate using the voices of others, the ultimate economy of words.

By appropriating and then displaying someone else's tweet, John could communicate to the original poster and to their followers, "Me too," in a heteroglossic way (Bakhtin, 1982, 1986). In addition, this grabbing and using of another's tweet was a means of instigating a conversation with that person. boyd, Golder, and Lotan (2010) explored the "conversational aspects" of retweeting, and similarly found that, for some users, retweeting was a way to bring "new people into a particular thread, inviting them to engage without directly addressing them" (p. 1). As I found in the participants' retweeting practice, it was as much about demonstrating alignment with others as it was about communicating a message.

However, this practice of retweeting a mundane greeting such as "hi" or "good morning" is not widespread on Twitter. This was a practice that was particular to John's Web sphere (although it may be observable in other niche Web spheres). In moving away from the conventions of the larger Twittersphere in this way, John demonstrated a sense of his Web sphere as separate and as a space over which he has some control. In some ways he recognized the authoritative discursive mores in the space (e.g., he writes in English, his utterances remain within the 140-character limit, he seems to seek to create meaningful content), but at the same time his discourse in the space is imaginative, and "tightly interwoven with [his] own word" (Bakhtin, 1982, p. 345). Local users shape the Internet in this way. Individual practices can break from the designers' intended uses, and as people in local sites contribute to the networked collection of texts, they influence it. Local practices intersect with technologies to create these spaces (Gordon & de Souza e Silva, 2011).

In one practice that drew on digital tools in inventive ways, John used a tool within Twitter for expressing agreement with or appreciation for a post, the "favoriting" feature, for curation. Favoriting a tweet shows the original poster that you liked, agreed with, or otherwise appreciated the post. Favoriting a post also links that post to the favoriter's Favorites section, allowing him or her to recall the tweet later by clicking on the Favorites tab. When I asked John about his use of the favoriting feature, he told me it was so he could read certain tweets later and have a "memory time:"

> When you favorite something…you can go back…it goes into like your little, like a little secret box like on the Internet or whatever. Just that only you can see, so if you wanna go look at it again. Or if you just goin, scrolling down through it, havin' like, a little memory time. That's why I favorite stuff. To see it again.

John's description suggests storage ("it goes into…a little secret box") of the past ("memory time"). Considering the Internet's rapid pace of information exchange, it is no wonder he leveraged the features of technology that allowed him to collect and save the bits and bytes that he deemed valuable. John's suggestion is that if he fails to favorite a post, he will never see it again. He seems to understand Web content as slippery, fast moving, and nearly ephemeral, and his invented practice is a means of creating a space for storage and reflection at once. He drew from the Twitter posts in his Web sphere to create this patchworked, multivocal new text that represented his history and would allow for future reflection. In this way, John saw his digital literacy practice as a means of not only connecting with others, but also a means of dialogue with the self.

Twitter has suggested users employ favoriting as a "curation model" since it allows for bookmarking, but neither of the other participants used favoriting as a curation tool the way John did. For example, rather than using favoriting as a means of curation, Aaliyah used the practice of favoriting a post on Twitter as a way of communicating with another person through the touch of a button. To favorite a post, she would tap the star icon under the post. This was a means of communicating with her social connections about her way of being in the world, and in that sense, was a composing practice that represented communicative practice but also a social self in flux.

Evolving Curation Mechanisms on Mobile Applications

When Twitter launched in 2006, there was no built-in curation mechanism. The first retweet was accomplished by hand; that is to say, a user simply typed "re tweet" and then copied and pasted a prior tweet from another user. Incidentally, the hashtag and the @-mention share that origin. They were started by users and then absorbed into the platform's architecture after platform designers took notice of user needs.

Similarly, Instagram did not have a built-in curation mechanism and to compensate, some users have taken up third party applications to supplement Instagram's native functionality. This work-around has not been popular largely because it is cumbersome to have to use two applications in tandem, and because Instagram has made many updates to its application program interface that interrupts the ways that these third party apps are able to communicate with Instagram's software.

Users working within the native features of Instagram's architecture who have sought to curate content have done so through tagging or using an @-mention. By using an @-mention in the comments section of an Instagram posting, users can effectively mark and send that content to a user. Instagram reported in 2015 that 40% of comments to Instagram photos included an @-mention. In response to this user-generated practice, Instagram added a new feature to their existing direct messaging feature, Instagram Direct, which allows for content curation. Specifically, the platform added an arrow next to the like and comment icons found under Instagram posts. A user can tap the arrow in order to curate that content by sending it to a friend or group. Content can also be curated through the sending of hashtag pages and location pages by tapping an arrow icon on the pages.

With the increasingly popularity of the practice of curation has come a variety of mobile phone apps designed to support users in curating content or to provide curated content to users. Apps that curate content according to users' preferences draw on the same principles that gave rise to Real Simple Syndication, or RSS. RSS feeds were widely used in the 1990s and early 2000s to allow users to subscribe to updates for their favorite websites. In an age of information overload, RSS allowed for information centralization. New apps like Curiyo, Feed.ly, and Flipboard accomplish this same kind of gathering and centralizing of information. This kind of customization has, in many ways, come to characterize postmodern consumer culture.

Conclusion

This chapter has explored the ways in which curation is a practice that is easy to engage in with the mobile phone because so much rich communication can happen with the touch of a button. The retweeting function in Twitter allows composers to speak through others' voices with one tap on the phone screen (much easier than typing a long-form text with the ungainly digital keyboard). Such communication is layered: it allows the youth to show agreement with the idea as well as to signal alignment with or to affirm the original poster. In the practice of text curation, the acts of reading and composing are almost collapsed. Through curation, that is to say, composers can read a text in their feed and then "grab" it and integrated it into their feeds. The roles of author and audience are blurred. Composers are in dialogue with the texts they consume in this way as they subscribe to and then take in posts from their connections, reflect on what they believe and disbelieve from this ideological world, and grab and display what they agree with or like in their own digital spaces, often communicating in direct and indirect ways with their connections. Youth also use curation (often through favoriting) as a means of holding onto digital content they want to revisit later, allowing for future reflection. Far from being a practice of usurping or recycling existing content, curation is a sophisticated and rich practice.

A Bakhtinian (1982) tracing of the phrase digital curation would acknowledge that this word is appropriated from museum or art gallery discourses. A museum or gallery curator is a specially trained individual who selects, acquires, cares for, and shares culturally significant objects like historical pieces or artwork. The term suggests elite high culture, as well as a sense of responsibility (because the word curator comes from the Latin curare meaning "to take care"). The phrase digital curation represents a pop-culture appropriation of a term once reserved for the specially trained elite. It also suggests a sense of the value that users of the term place on digital artifacts.

Although Bakhtin (1986) wrote before the age of networked computing, his work seems to prefigure the massive, World Wide Web of voices and linkages between voices that indicate texts in dialogue with each other. From a Bakhtinian (1986) perspective, the retweet and other forms of digital curation are heteroglossia writ large: Aaliyah, like John and Eli, appropriated and thus spoke through the voices of others. They grabbed existing utterances and populated them. The retweet is, to paraphrase Bakhtin (1986), half ours, and half someone else's. This suggests a cultural model that regards communication as a form of connection. As two voices unite, they are both louder or more resonant. In the fast-paced discursive environment that is Twitter, where utterances are rapidly buried by the influx of new utterances, this approach is quite prudent.

In the participants' digital composing practice, they were able to assign value to texts. They were in control. As the Web moves from Web 2.0 to Web 3.0, some experts predict that subscription and syndication, personalization, and increased control over the data to which people have access through the Web will characterize the space of the Web. The data collected in this research effort suggests new approaches to composing in this social space of the Web are characterized in much the same way: students subscribe to others' posts through "following," consume and then assign value to texts, capture texts, and manipulate or distribute texts as a composing practice.

Even for youth who compose primarily by way of mobile phones with their limited affordances for digital composing, meaningful cultural production is afforded by digital curation as a means of composing. The phone as a point of access and a tool for composing did not prevent the participants from interacting in the digital space. Some might criticize the act of digital curation as a means of composing because it is not creating something new, but instead responding to something already public. Curation is not a practice commonly valued in schooled literacy spaces, for example. However, I see the practice as responsive to or in dialogue with the current moment (Bakhtin, 1982). There is much ready-made content for youth to find and use to express themselves without having to create an altogether new text and this is the reality of remix culture (Jenkins, 1992, 2006a, 2006b).

All language is heteroglossic according to Bakhtin (1982), but only some is polyphonic, containing many voices brought together within the authorial word. The Twitter feed of all three participants were polyphonic (Bakhtin, 1982). They brought in the voices of others without commenting directly on them through the curation practice. However, as the account holder or creator, the respective participants held the authorial power. Their voices were dominant in their own Web spheres and assert a sense of truth in relation to the diverse voices in that space. In some ways, the dominance of their authorial voices was challenged, however. In digital text, the reader is afforded the opportunity to choose how to experience the text. One can navigate around the page autonomously to experience a number of conversations and thus the authorial voice loses dominance to some degree.

These practices and their related ethos have, I argue, implications for schooled literacy learning. All three of the participants saw both connections and disconnections between their schooled and digital literacy practices. They did not dichotomize them, and called upon schooled literacy skills at home and home-fostered literacy skills at school. Eli and Aaliyah first started using digital media and going online at home, while John initially learned these skills at school and brought them home. The participants' literacy practices and skills were not bounded or relegated to particular spaces; they called upon all they had in their repertoire as needed to be successful across spaces. The participants' cultural models were informed by their interactions in online spaces. Sometimes they took up widespread and dominant practices in those spaces (Bakhtin's authoritative discourse), and sometimes they created and contributed their own, more localized practices (internally persuasive discourse; Bakhtin, 1982). For example, Eli described eschewing some widespread practices, such as communicating with just one person through public tweets. John created his own practice of using "favoriting" to create a space to reminisce, and Aaliyah favorited tweets in order to show agreement.

Considering the popularity of the practice of digital curation, I argue that youth who engage in this practice are being socialized into a space with a philosophy of language that is in contention with that of schooled literacy. Whereas the participants in this study borrowed and remixed texts without attribution as normal practice, in school such a practice would be considered plagiarism. Whereas coopting the voices of others is a legitimate way of participating in the digital space, where digital curation is popular practice, in school, taking up the voices of others to express one's ideas is seen as plagiarism or academic dishonesty. Youth would benefit from opportunities to critically analyze the perspectives of other thinkers and to speak through the voices of others as they work to assemble their emerging identities while learning about academic citation conventions and mobilizing their voices for participation in society.

References

Bakhtin, M. (1982). *The dialogic imagination*. Austin, TX: University of Texas Press.

Bakhtin, M. (1986). *Speech genres and other late essays* (V. W. McGee, Trans.). Austin, TX: University of Texas Press.

Bourdieu, P., & Passeron, J.-C. (1990). *Reproduction in education, society and culture*. London: Sage.

boyd, d., Golder, S., & Lotan, G. (2010, January). *Tweet, tweet, retweet: Conversational aspects of retweeting on Twitter*. Paper presented at the proceedings of HICSS-43 IEEE Computer Society, Kauai, HI.

Brenner, J. (2013). *Pew Internet: Social networking (full detail)*. Retrieved from http://pewinternet.org/Commentary/2012/March/Pew-Internet-Social-Networking-full-detail.aspx

Gee, J. P. (1996). *Social linguistics and literacies: Ideology in discourse* (2nd ed.). New York, NY: Falmer Press.

Gordon, E., & de Souza e Silva, A. (2011). Globalization. In E. Gordon & A. de Souza e Silva (Eds.), *Net locality: Why location matters in a networked world* (pp. 168–183). Chichester, West Sussex: Wiley-Blackwell.

Jenkins, H. (1992). *Textual poachers: Television, fans, and participatory culture*. New York, NY: Routledge.

Jenkins, H. (2006a). *Convergence culture: Where old and new media collide*. New York, NY: New York University Press.

Jenkins, H. (2006b). *Fans, bloggers, and gamers: Exploring participatory culture*. New York, NY: New York University Press.

Milner, R. M. (2013). Hacking the social: Internet memes, identity antagonism, and the logic of lulz. *The Fibreculture Journal, 22*, 61–91.

· 8 ·

CRITICAL DIGITAL LITERACIES

Abstract

This chapter argues the need for a new conceptualization of critical digital literacies in an age where so much youth composing happens in networked digital contexts with and through the mobile phone. Through illustrative data collected over the course of this study, this chapter shows how youth take up marketing discourses used in advertising in digital spaces in their digital composing practice. The chapter closes with a discussion of the need for a kind of critical digital literacy that accounts for reading digital spaces in addition to digital texts especially in an age of curation, meme-sharing, and identity construction with and through digital texts.

Keywords: Critical literacies, critical digital literacies, marketing discourses, advertising, identity, curation, memes, mobile phone, digital spaces

> *This classroom has "Read 180" stuff everywhere. This is a Scholastic moneymaker. A large poster hangs, "This is a Read 180 Classroom!" "Read 180 Software Station." "Read 180 Level 1 Books." "Read 180 Instructional Model." "This is apparently a derivative of AR (Accelerated Reader)." Some of the Read 180 books are audiobooks. They are mostly YAL (Young Adult Lit). That hardly existed when I was teaching five years ago. A poster on the wall near all the Read 180 Stuff:*

> *Stay focused on reading your book daily.*
> *Complete your reading log entry for the day.*
> *Complete the quick write as you come to it while reading.*
> *When finished reading book, complete the Graphic Organizer Worksheet.*
> *Bring your folder and completed work to small group for teacher check/approval.*
> *Take the Scholastic Reading Counts quiz on the computer. (field notes, 11/5)*

As the field notes from my first day of participant observation in Ms. Harrison's classroom illustrate, the initial thing I noticed in her classroom was this corporate presence. Everything was branded; books, posters, tests, and even teacher-made materials carried the trademark of a manufacturer.

The fact that I noticed this right away was not because branding, marketing, and commercialism in the spaces for education were particularly new to me. During my time in the classroom as the teacher of record, a required element of all instruction was something involving Accelerated Reader (AR). AR is a product similar to Read 180 sold by a company called Renaissance Learning that organizes, monitors, and ultimately monetizes the practice of reading. In the school where I taught, all English teachers were required to offer the first 15 minutes of class time as AR reading time. Of course, this meant we also had to make books for AR reading available to students. Unfortunately, students only earn "AR Points" by taking AR tests on the computer (available for purchase to schools from Renaissance Learning, a major education company in the United States). Only certain books have AR tests, however. This meant that the students' AR reading was restricted (and resultantly, dictated by that company). In other words, Renaissance Learning had a lot of power to influence what my and other students read.

The students were keenly aware of what books would afford them AR points and which ones would not. I became aware of how incentivized the students were by AR points and how this incentive had a real bearing on their literacy early on. Often I would notice a student who, rather than reading, was staring at the ceiling or off into space during AR reading time. Selecting a book off my shelf that I knew would appeal to that student, I would hand it to the student so he or she could harness the fifteen minutes of reading time. More than once, I was met with the reply, "But there's no AR test for this, Miss Warner." Accordingly, I would often hear my colleagues lament the fact that Accelerated Reader was squashing the students' love of reading for reading's sake.

In doing some research into Read 180, I found that it is a "Reading Intervention Software Program" aligned with the Common Core State Standards, an educational initiative in the United States that includes English language arts standards created by Hasselbring and Goin (2004) at Vanderbilt University and advertised and sold by publishing house Scholastic. The name "Read 180" is intended to suggest that the program goes beyond AR's simple testing and points system; Read 180 combines the computer-assisted tests similar to those of AR with other instructional approaches such as in-class group instruction and computer activities. Although I did not see Ms. Harrison use the program during my observations, in the classroom space, Read 180 was a very-visible corporate presence.

Schools expose students to companies and their brands and teachers become complicit in promoting those brands when schools require corporate programs like Accelerated Reader or Read 180. Private industry is present in United States classrooms beyond simple reading programs. Textbooks, curricula, professional development, and teaching materials created by corporate entities are present across the subject areas in public schools. A commercial agenda is thus inextricable from teaching and learning and commercial entities are afforded the power to influence what students learn, when they learn it, and how they learn it.

This chapter explores the ways that a critical literacies framework that promotes active analysis to uncover structures of power can be applied to interrogate not only the physical spaces and texts involved with literacy learning but also digital spaces and texts. As this chapter will make plain, as youth spend more time online than ever before and as marketers move into the spaces that youth inhabit, digital literacies education in an age of ubiquitous mobile phone-based literacy practice must be engaged from a critical perspective in order to uncover such underlying power structures. However, this critical perspective, as I'll explain, must be reconceptualized to make room for the kinds of new literacies encouraged by affordances of the mobile phone in particular.

Critical Literacies

Sociologists have examined the ways that, through the organization of their physical and social spaces, schools reproduce social and cultural status (Bourdieu & Passeron, 1990). Using a framework that regards literacies as social practices can help to foreground the ways that power organizes the physical and social spaces that youth inhabit. Critical literacies frameworks have been taken up by researchers interested in teaching youth to interrogate texts, which contain and represent cultural information, to uncover issues of social injustice and to critique the status quo (Freebody & Luke, 1990). More recently, the concept of critical digital literacies has been applied in studies of youth digital literacy practice to highlight shifts in disposition and identity. Most often, this research uses digital texts as the unit of analysis and regards the digital texts in much the same ways that physical, print texts were regarded in early critical literacy work; critical literacies frameworks and concepts are superimposed/overlaid on digital contexts. I argue in this chapter that when researchers and practitioners talk about critical digital literacy, we need to go beyond text-based approaches to talk about making the circulation of power in the digital spaces that youth inhabit legible.

Corporate Presence in Digital Spaces

During my ethnographic work in the physical spaces of schooled literacy such as within the walls of White Birch High School, I noted the many ways in which corporations and other commercial entities dominated the physical spaces youth inhabit. In the digital spaces, this was unfortunately also the reality. Throughout my observations, I took note of the myriad ways in which private, commercial entities showed up in the digital spaces inhabited by the participants.

For this reason, I wasn't particularly surprised about the ways the participants integrated commercial content in their postings. The influence of corporate interests is palpable online with advertisements on the margins of almost any site one might visit. Music videos, which John and the other participants watched online at YouTube, were not only preceded by commercials, but also featured prominent product placement. One of all three of the participants' favorite artists, rapper 2Chainz, is known for the copious product placement in his music videos. His music videos were one of the texts they consumed as they interacted online. Contemporary youth marketing mobilizes means of deep, long-term identification (e.g., the use of social media accounts of celebrities) to advertise to youth consumers. Mobile phone-compatible social networks popular with youth like Twitter, Instagram, and Snapchat have monetized their sites by incorporating advertisements into the sites. The advertisements take the form of what look like normal posts and can easily be mistaken for organic social content.

On Instagram, advertisements are only differentiated from social content by the small word "sponsored" where a post's timestamp normally appears. But outside of Instagram's direct sponsors, product placement and subtle advertising was common, too, in the participants' Web spheres. In one artifact collected in the course of this research, a post by Beyonce (a popular musician whom some of the participants followed), the artist subtly pushes Pepsi in a four-pane Andy Warhol-inspired graphic showing Beyonce's face in three panes and a can of Pepsi in another. This advertisement appeared among her Instagram photos of her everyday life and was not immediately recognizable as an advertisement since it employed an aesthetic aligned with Beyonce's hip aesthetic. Similarly, in another artifact, LeBron James, whom some of the participants followed, posted a photo of his sneakers with the hashtag #AIRPYTHON to let his followers know the name of the shoe he is promoting. In these cases, the line between authentic social media interactions and paid advertising blurs.

Benjamin, Dhalwal, Brown, and Kunto (2011) find that contemporary youth marketing mobilizes means of deep, long-term identification like using the social media accounts of celebrities whom teens admire because the trust and attention of the modern young person are hard-won. Livingston (2010) says teens are "immune" to traditional advertising and so advertisers have looked to native advertising to reach young consumers. In engaging a dialogic lens, I noted the ways that the discursive and multimodal features of such native advertising was taken up by the participants who composed in these advertisement-heavy spaces for composing.

For example, the youth in this study demonstrated relationships with brands in their social media postings in a variety of ways. At times they might feature a logo in a posting, or show off a new pair of designer shoes. Other times, though, the commercial influence showed up in highly complex ways. For example, in Figure 8.8, John references his Ralph Lauren designer clothing in the linguistic text, in the caption "polo hat, polo tee, polo all ova me." He creates his own jingle for Ralph Lauren, in effect offering free advertising to the company. Through the visual and linguistic modes he employed here, John not only demonstrates that he is conversant with branding and commercialized discourses, but he becomes complicit in a company's marketing objectives.

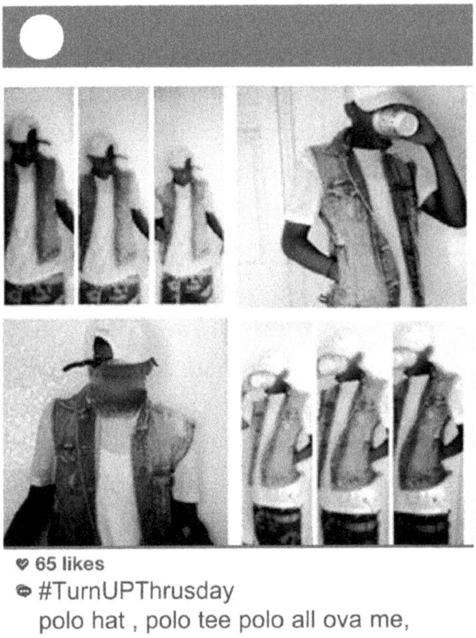

Figure 8.8: John Showing off His Ralph Lauren Polo Designer Style.
Source: Author.

All of this is not to say that such a posting is evidence that John lacks agency. In fact, Figure 8.8 shows that John takes up youth marketing discourses of the online marketing and advertising industry as he works to assemble and create his identity using digital media. In fact, much of the data I collected similarly bore out a collision of creative visual composing, identity work, and commercial influence.

For example, when I probed John about his posting of himself decked out head to toe in Ralph Lauren Polo, and his caption, Polo Hat, Polo Tee, Polo all ova me, he made clear to me that he saw his engagement as reflective of his fashion sense, which he saw as a core component of his identity. However, John used a representational palette that was comprised of private, commercial interests by posting to social media giant Instagram which depends on user generated content like John's to exist and by providing the Ralph Lauren company with free advertising through modeling their clothing, making up the jingle Polo Hat, Polo Tee, Polo All Ova Me (Polo being a Ralph Lauren brand), and by disseminating the photo to his hundreds of followers. I argue that the obfuscation of advertising in the spaces where the participants interacted, including school, where I noted the kinds of ubiquitous advertising that Kohn and Shannon (2002) write about, and the texts they read (e.g., product placement in music videos, like those of the participants' favorite rapper 2Chainz who has been widely criticized for "selling out" to advertisers) manifested in the participants' discourse.

Furthermore, in taking up and circulating these discourses, the content that the participants in this study create has economic value, none of which is afforded the participants, who were unaware of these structures. They saw themselves as agentic creators of their own identities even though they pulled from a representational palette that was strategically presented them by private commercial entities to do so. As such, youth writing in networked publics face myriad challenges that they may not be equipped to face.

Shout Outs, Followers, and the Commercialization of Friends

Not only were marketing discourses present in youth postings with logos and marketing jingles but I also found marketing and commercial discourses in youth attempts to "promote" friends. For example, John often posted photos of friends, which he found on the friends' pages and reposted, with captions such as, "S/O [shout out] to @(username) follow dis guh," "Everyone stopp what youree doinn andd foolow meh now @(username) @(username) @(username),"Everyone go follow dys girl @(username)." A shout out is an acknowledgement, and in this case, it is a promotion of the friend. By posting to his followers, John increases her exposure. He promotes her and demands that others follow her in a way that positions her as an entity in need of publicity. Such discourse positions friends as commodities rather than people. The participants in this study subtly and overtly advertised brands, companies, and even their own friends in their online writing and took up discourses of commercialism.

Buckingham (2008) identified a shift from youth media consumption to production. Much of the data collected over the course of this study, however, suggests a lamination of consumption and production. As youth take up commercialized discourses in their media production, they contribute to capitalist systems while also remixing capitalist discourses. Youth are writing their worlds and their selves into being with digital media in a space where learning to do so happens socially, by observing the discourses present in the space, discourses laden with commercial interests not often visible to them.

Identity Play in Corporate-Owned Spaces

Whether or not youth identity play with commercial discourses represents an assertion of their own power and agency, it is necessary to also critically examine the spaces in which youth text production happen. I came to this realization as I mapped the spaces and circulation of power related to youth digital text production over the course of data collection. I began to see some aspects of the participants' digital composing practices within commercially-owned spaces such as Instagram and Twitter by way of their mobile phones which could prove to be problematic for them in the future given the nature of digital texts with their replicability and permanency. It may be the case that young people are experimenting with identity in corporate-owned spaces and thus losing control of or power over some part of their identities to private industry. To illustrate, I can point to Eli's digital composing practice. As I carried out this research, I struggled with issues of representation with regard to Eli. Early on, a colleague responding to the work warned me that I needed to be "careful with how I was representing" my participants. This was in response to some data wherein Eli was demonstrating gang affiliation through social media postings; he often posted photographs that expressed affiliation with a neighborhood gang. Because he was frequently in trouble in the schooled space, it was not surprising to me that he acted out a "bad boy" narrative online. The way he was positioned at school was reflected in his digital literacy practice. Moje (2000) asserted that teens shut out from the dominant "story" because of ethnicity, race, or social class used gang-related literacy practices to write "themselves into the world" (p. 652). Literacy practices involving gang practices "may serve as tools for transforming thought and experience in the lives of marginalized youth" (Moje, 2000, p. 652) and they certainly seemed to for Eli.

I recognized Eli's digital gang affiliation practices were a form of agency, of exploring an identity and of making himself known. What I was seeing was a young person looking to fit in and who was acting out a "bad boy" persona likely because it gave him a sense of power and because it was the way he was positioned in the schooled space (Ferguson, 2001). However, I knew that to an outsider, it might appear that his engagement was negative or that what one could see online was an accurate reflection of who Eli was as a person. I knew that in actuality he was venting and performing. Ultimately, I chose to represent accurately what he did "warts and all" and to obscure his face to protect his identity because I believe it is important to include even digital composing practices that might seem "negative." I the end, I think this inclusion is important because although digital texts feel abundant, ephemeral, and hidden among the digital deluge to youth, Eli's identity work on social media became the property of a profit-seeking company. As youth are using media to write themselves into the world, they are, in effect, leaving a permanent record of youthful indiscretions that can constrain choices in their social futures (Kinzer, 2010). All of this is problematic in that digital content never goes away. As Eli carried out identity play without much thought in a digital space that to him felt intimate and safe, he did so without the awareness of the repercussions of such play, of composing multimodally with his physical self, his body, in a space that is permanent and over which he had ultimately little control, since his postings were in a corporate-owned space.

This case highlights the need for digital critical literacy instruction as part of multimodal composing instruction for the 21st century. Whereas in the past, youth could experiment with identity though clothing at school, posters on a bedroom wall, or other means that persist only in people's memories, youth today use digital media to do so and thus leave a record of such identity play. Youth need a safe space to grapple with ideas and experiment with identity. without going on the public record. In addition, digital spaces should be made legible for youth so that they have some control over their data. In other words, youth need explicit, direct instruction around the nature of digital composing environments, especially those commonly accessed through mobile phones. Social media platforms such as Facebook have traditionally made it difficult to understand how to control one's data through convoluted and overly complicated privacy settings.

Curation and Critical Digital Literacy

As discussed in previous chapters, not all of the texts the participants in this study dealt with were original texts. Some of the composing engaged in by the participants in this study consisted of circulating texts that existed in the spaces they inhabited. For example, all three of the participants in this study spread and remixed Internet memes, highly spreadable, digital cultural artifacts. An Internet meme is an idea, image, video, or other piece of digital cultural information that is transmitted and often imitated. Lankshear and Knobel (2006) defined memes as,

> contagious patterns of "cultural information" that get passed from mind to mind and directly generate and shape the mindsets and significant forms of behavior and actions of a social group. (p. 199)

Evolutionary biologist Dawkins coined the term meme in 1976 to describe the transmission of cultural information that spreads:

> Examples of memes are tunes, ideas, catch-phrases, clothes fashions, ways of making pots or of building arches. Just as genes propagate themselves in the gene pool by leaping from body to body via sperms or eggs, so memes propagate themselves in the meme pool by leaping from brain to brain, via a process which, in the broad sense, can be called imitation. (Dawkins, 1976, p. 192)

Lankshear and Knobel (2006) pointed out that Internet memes do not have the same kind of sticking power as the cultural information of which Dawkins wrote, but the method of transmission is similar. Lankshear and Knobel (2003) called memes a "new" literacy.

One meme collected over the course of the research was one that John retweeted. The meme demonstrated intertextuality in that it melded of the image of SpongeBob Squarepants and a separate meme, "hoes be like." The still used as the basis of the meme is taken from the SpongeBob cartoon and becomes the visual template waiting for text overlay. "Hoes be like" was a viral, memetic phrase, commonly used as a preface to negative stereotypes of women on the Internet. John communicated through his circulation of this meme misogynist, patriarchal, antifeminist views, and makes them more readily available for further spread, encoded in this Internet meme. Most negative memes are imbued with or are ruled by "the logic of the lulz," lulz being a derivative of the shorthand LOL or "laughing out loud" (Phillips, 2012, as cited in Milner, 2013). This is "the ironic critical logic [which] often antagonizes the core identity categories of race and gender, essentializing marginalized others" (Milner, 2013, p. 63). Memes like John's SpongeBob meme reappropriate pop or subculture images, matching them with a message ruled by the logic of the lulz and employing "incongruity" (Shifman, 2011, as cited in Milner, 2013) to generate humor. Indeed, when I asked John about the image, he said, "It's just funny." He did not sense that the message was particularly offensive, but seemed instead to focus on the incongruity between the SpongeBob cartoon and the particular cultural critique with which the image was paired.

In the early days of the Internet, it was predicted that because communication was mediated and thus disembodied, race, gender, and other differences would be irrelevant or disappear. Instead, people use the Internet to make sense of their racial and gendered identities and form community based on these identities online (Byrne, 2008a, 2008b; Everett, 2004; Nakamura & Chow-White, 2011). The idealized, race-free version of the Web sprung from the early text-based Web; with the multimodal Web, Nakamura (2008) maintained that people represent and play with race and other markers of identity more than they did previously.

While John sometimes circulated harmful ideologies as we see with the Spongebob meme example, he also often disrupted the circulation of harmful ideologies through his productive subversion. For example, in one interview, he suggested that he might use the phrase "Team Nappy Head" on his profile as an identifier. "Team Nappy Head" represents a reclamation of a phrase that was used historically as a negative way to describe curly or textured hair. Through this reclamation, John asserts pride in his racial identification as African American. He also uses this as a way of finding and aligning with others with the same point of view about this phrase (pride instead of shame) as a "team." At the same time, it seemed at times John did not seem to recognize the harmful ideologies he circulated (e.g., "hoes"). Helping students to recognize the circulation of discourses should be part of critical literacy practice for the 21st century, particularly in a composing space with an ethos of curation where cultural messages are encoded in texts that youth distribute with rapidity.

As youth compose by curating images, video, and text, they heteroglossically take up prior discourses, even if such texts are recontextualized and are coopted by youth for their own purposes. As youth actively engage with, rework, and appropriate ideological messages in texts, particularly within processes that are characterized by their rapidity (curation, microblogging, or social photosharing) this need for critical literacies becomes apparent. For example, the participants took up discourses of private commercial interests in creating their online profiles with digital media. Critical literacy practice for the 21st century should thus offer a way to teach young people to recognize this new advertising as part of critical literacy. In addition, youth need to understand that in taking up the semiotic and discursive affordances in a space and circulating them they are in effect brokers of ideologies. This has to be a part of learning to be literate in the digital age. Curriculum that is pliable and responsive rather than static should be the new imperative.

Reading New Social Spaces as Digital Literacy

Not only should critical digital literacy teach youth to interrogate the texts they read and circulate, it should also involve teaching youth how to read digital spaces. Digital spaces have different mechanisms by which youth can be surveiled, have different structures for user control, and exist within particular social contexts. Different spaces offer different affordances for agentic action. For example, while Eli's composing practice was a mobile practice, Deleuze and Guattari (1987) reminded us that mobility is channeled, that even the nomad moves along routes often provided by conduits in space. As Frith (2012) said, "how much control individuals have over their experiences while mobile is a function of how legible particular spaces are for the people who move through them" (p. 134). In some cases, the spaces Eli moved across did not seem to be legible for him. I noted the ways dangerous spaces were created after I found from Ms. Harrison that Eli was expelled from White Birch High School for fighting. She informed me that the argument began as a comment on one of Eli's Instagram photos, and escalated at school. I was able to find the photo. It was a photo Eli had posted of himself from his mother's car, and the comment thread attached to the picture contained bullying discourse.

Antagonist:	Soo you cool in shit?
Eli:	Hell ya nigga! Haha
Antagonist:	Or nahh
Eli:	U seem 2 6e a hater 6itch
Antagonist:	Lls [laughing like shit] bitch [includes emojis, icons only visible to iPhone users] but you can't see those emojis huh ?
Eli:	6itch I gotta app dat let me see em so suck ma dik nigga!
Antagonist:	Boy fuck you boy!, but fuck you and that app shit !
Eli:	Pleas get ur hatin azz off ma page. Ain't no6ody got time for diz
Antagonist:	you sound like ahh bitch ! [Emoji of a laughing face and a peace sign]
Eli:	6itch kan u pleaz get a life n get off ma instahram 6eforei 6eat u razz
Antagonist:	Aight bitch I know you seen the peace sign ! I'm tired of fucking with you ! But beat who asss? Lls
Eli:	Urz 6itch. Square da fuk up.
Antagonist:	Wassssssuppp though ? Lls ima come mcdonalds too fight you, get yo ass fired ! [emoji of an ok sign]
Eli:	Wrd 6itch I wrk 3:30–11:00 ill 6e waitin 6itch
Antagonist:	Yeeeeen gotta clock in cuhh ! Ima be there waiting ! Ima fuck some shit up in there too !
Eli:	Hahahahahaha ighht nigga! :) see u dare!

The antagonist never met Eli at his work; however, the next time the boys saw each other at school, they fought. Eli was expelled from White Birch High School because it was not his first disciplinary violation.

My analysis of space in relation to this particular text, for example, highlighted the way an antagonistic youth sought out the space of Eli's Instagram, in particular a photograph of his face, to carry out his attack. This space was legible to the antagonist as a safe space for him to do so, unnoticed by authority figures. In addition, I identified this space to be a public space (in contrast to private spaces such as private messaging, text messaging, or other peer-to-peer methods of communication) where he could gain the attention he sought from peers. Although the space Eli created through the creation of and continued engagement with Instagram may have been one he experienced as "safe," as it was comprised primarily of friends, it was a public and penetrable space, and this fact had very real consequences for him, and altered his life trajectory. It was a youth space not monitored by authority figures who could regulate behavior or prevent bullying. Ultimately, this led to Eli's school failure. He needed more mechanisms to regulate the social space. These kinds of mechanisms are in place in the material space of the school, but not the online social space.

The antagonist had information about where Eli worked as well as how to locate Eli through his photographs and the mapping function of Instagram. Although he created the space, Eli was ultimately unable to regulate it. He read the space as a safe space for peer-to-peer discourse, when in actuality an interaction in the space led to his school failure. He needed instruction about audience in digital writing contexts and about the geographic and temporal scalability as well as the persistence of digital texts in a context of transnational global flows (Castells, 1996). After I found out about his expulsion, Ms. Harrison informed me that she had, just before the incident, recommended Eli for honors English for sophomore year. He would now miss out on that opportunity and the others that would go along with continued enrollment at White Birch High School.

Not long after he was expelled, Eli posted a selfie to his Instagram in which he donned his White Birch High sweatshirt, and in the picture he held up the school initials. He wrote as the caption: "U can kick me out 6but u Kant take my skool spirit away" with an emoji of the number 100 (a reference to the phrase, "keep it 100," which means staying loyal). I knew that he would pursue a new route and later saw on Instagram postings from his life at his new school, but I could not help but think about the ways the lives of youth in this new connected world may be subjected to more unpredictability than ever before given the myriad of digital spaces they inhabit but cannot fully read.

Critical Digital Literacy and Agency: Front End *versus* Back End Spaces

Throughout the study, I took note of not only the spaces the participants created and frequented but also those they did not. All of the participants remained on the front end of digital spaces. In the field of computer science, the term front end refers to the part of digital media that one can see and interact with. It is usually designed to take in input. Back end, on the other hand, refers to the code that supports the front end. The back end is hidden from users, and so it is not particularly surprising that the participants did not frequent back end spaces. However, there are some implications to their remaining in the front end interface. For example, to some degree, they are limited in their actions by what those on the backend design. Their activity is channeled in particular ways by back end design to some extent. Even when they do produce content, it is embroiled within the desires and plans of profit-seeking corporations that rely on consumer-produced content as in Web 2.0. (As described in the introduction to this chapter, school was not safe from such interests, either. Commercial interests permeated the schooled space as well. From the corporate-controlled tests that students must take as the "official" measure of their literacy to the books they read as part of programs like Read 180, this influence was ubiquitous). Curriculum that responds to the rapid privatization of both physical and digital spaces is necessary to position youth as agentic.

Adolescent learners interact primarily on the front end of computers (what they see on the screen) and many do not have any exposure to what is on the back end (computer code). Many do not know how the technology works, thus their digital literacy is severely limited. The recent widespread interest in coding or computer programming as evidenced by The Hour of Code Initiative where teachers were challenged to teach coding for one hour to students, which was wildly popular in the United States, is proof that this is changing. There are a multitude of platforms for coding that are free to educators and students and are intuitive and user-friendly and could be used to provide youth access to that back end space. Computer programming allows students not only to learn how to use computers, but how computers work. Without the knowledge of how computers work, students are positioned as consumers only. Becoming a producer is a vital 21st century skill. As more and more of their worlds are mediated by computers, it will benefit students to increase their awareness of how these systems work and is, I argue, the only in-road to a truly critical digital literacy.

Conclusion

With the widespread adoption of mobile phones, the corporate influence on literacy practice now extends far beyond private industry owned course materials or curriculum. Corporate influence in spaces for literacy bleeds across the spaces of school, home, and across digital spaces for composing. Young people know how to block or otherwise avoid commercials in their media consumption practices and even online-only commercials are, marketers are finding, ineffective at reaching younger generations. Since the cellphone is a ubiquitous companion among this demographic, it is easy to see why marketers are flocking to mobile-phone-compatible advertising.

Yet, most young people are not afforded any education around mobile phone-based literacy, including critical digital literacy to help them understand the ways in which the hardware, software, and social practices position them in the world as producers, consumers, or somewhere in between. Instead, they learn through engagement in these (often corporate-owned) spaces and through interacting with other users in those spaces who are often as inexperienced at decoding discursive messages and identifying the circulation of power as they are.

Since the youth in this study were learning to communicate online not through classroom instruction or any other form of direct instruction but through interacting in online spaces, observing the postings of other people in those spaces, and taking note of the social feedback they received on their own postings, it made sense that they took up and circulated the discourses present in the spaces they inhabited. In learning socially how to compose in these spaces with heavy commercial presence, they thus unwittingly take up and use private commercial discourse to represent their identities.

Bakhtin tells us that our words come from other people's mouths; this is what he calls polyphony. When youth use the words from the "mouths" of corporations to construct their identities, they come to define themselves as consumers. If we see education and in particular literacy education as a practice of freedom, we must consider how enmeshed in commercialized structures youth are when they enact literacy in online spaces.

Since the participants took up and circulated the discourses in the spaces where they engaged, they were limited in their expression by the types of available discourses. This is salient since youth develop their approaches to writing by taking up the forms and conventions of discourses in the spaces where they interact. If the only available discourses in a particular digital space are in some way deleterious, youth may end up engaged in the recirculation of negative discourses.

While early definitions of digital literacy focused on computer skills and technological competencies alone were narrow and later notions of critical digital literacy that considered analytical viewpoints and social competencies broadened early definitions in important ways, what has been done in this area to date has been insufficient. To be prepared to be engaged and ethical actors in the digital space, youth need deep technological skills that move their action from the front of the screen as content consumers to behind as not only content but software creators. In this way, they can gain an understanding of the inner-workings of technologies in order to get a better sense of how their activities and practices are shaped by the hardware and software that they take up in their literacy practices. Knowing how to use technologies is one piece of the puzzle, but tinkering and making with software is necessary to gain a rich perspective on how technologies act on us in the world.

Critical readings of spaces and texts can complement software creation, giving youth the tools to understand how texts act in the world from the other side of the screen. The Internet and digital texts can be viewed as cultural artifacts that exist within sociocultural formations and also encode sociocultural messages. Interrogating the processes and conditions for digital text production, translating the cultural messages they contain, and exploring their effects on audiences or how they are otherwise taken up in the world is central to this piece of critical digital literacy. This marriage of front end and back end literacies is what will be needed for students to achieve true critical digital literacies in the twenty-first century.

References

Benjamin, F., Dhalwal, J., Brown, G., & Kunto, G. (2011). *The youth marketing handbook*. London: Mobile Youth.

Bourdieu, P., & Passeron, J.-C. (1990). *Reproduction in education, society and culture*. London: Sage.

Buckingham, D. (Ed.). (2008). *Youth, identity, and digital media*. Cambridge, MA: MIT Press.

Byrne, D. N. (2008a). The future of (the) "race": Identity, discourse, and the rise of computer-mediated public spheres. In Everett, A. (Ed.), *Learning race and ethnicity* (pp. 15–38). Cambridge, MA: The MIT Press.

Byrne, D. N. (2008b). Public discourse, community concerns, and civic engagement: Exploring Black social networking traditions on BlackPlanet.com. *Journal of Computer-Mediated Communication*, *13*, 319–340.

Castells, M. (1996). *The rise of the network society: The information age: Economy, society, and culture* (Vol. I). Oxford: Blackwell Publishers.

Dawkins, R. (1976). *The selfish gene*. Oxford: Oxford University Press.

Deleuze, G., & Guattari, F. (1987). *A thousand plateaus: Capitalism and schizophrenia* (B. Massumi, Trans.). Minneapolis, MN: University of Minnesota Press.

Everett, A. (2004). On cyberfeminism and cyberwomanism: High-tech mediations of feminism's discontents. *Signs*, *30*(1), 1278–1286.

Ferguson, A. A. (2001). *Bad boys: Public schools in the making of Black masculinity*. Ann Arbor, MI: University of Michigan Press.

Freebody, P., & Luke, A. (1990). "Literacies" programs: Debates and demands in cultural context. *Prospect*, *5*(7), 7–16.

Frith, J. (2012). Splintered space: Hybrid spaces and differential mobility. *Mobilities*, *7*, 131–149.

Hasselbring, T. S., & Goin, L. (2004). Literacy instruction for older struggling readers: What is the role of technology? *Reading & Writing Quarterly*, *20*, 123–144. doi:10.1080/10573560490262077

Kinzer, C. K. (2010). Considering literacy and policy in the context of digital environments. *Language Arts*, *88*, 51–61.

Kohn, A., & Shannon, P. (2002). *Education, Inc.* Portsmouth, NH: Heinemann.

Lankshear, C., & Knobel, M. (2003). *New literacies: Changing knowledge and classroom learning*. Philadelphia, PA: Open University Press.

Lankshear, C., & Knobel, M. (2006). *New literacies: Everyday practices and classroom learning* (2nd ed.). New York, NY: Open University Press.

Livingston, A. (2010). *Secrets of advertising to Gen Y consumers*. Vancouver, BC: Self-Counsel Press.

Milner, R. M. (2013). Hacking the social: Internet memes, identity antagonism, and the logic of lulz. *The Fibreculture Journal*, *22*, 61–91.

Moje, E. (2000). To be part of the story: The literacy practices of gangsta adolescents. *Teachers College Record*, *102*, 652–690.

Nakamura, L. (2008). *Digitizing race: Visual cultures of the Internet*. Minneapolis, MN: University of Minnesota Press.

Nakamura, L., & Chow-White, P. (2011). *Race after the Internet*. New York, NY: Routledge.

Phillips, W. (2012). The house that fox built: Anonymous, spectacle and cycles of amplification. *Television and New Media*, *14*(6), 494–509.

Shifman, L. (2011). Anatomy of a YouTube meme. *New Media & Society*, *14*(2), 187–203.

· 9 ·

CHRONOTOPIC EXPLORATIONS OF MOBILE PHONE COMPOSING

Abstract

This chapter explores mobile phone-based composing through the lens of Bakhtin's chronotope. Bakhtin (1981) borrowed from the theory of relativity the term chronotope (literally "time-space") to emphasize the inextricable nature of these two concepts. Rather than viewing time and place as the backdrop for interaction and communication among characters, he saw them as central and as indicative of an author's worldview. This perspective is useful in that it can help to challenge the apparent "naturalness" of literacy education as it currently exists in regard to time, space, and opportunity for learning in order to address associated pedagogical challenges. For Bakhtin, the novel (and other representations of events and lifeworlds) invokes time and space and particularly how people live in time and space and leave markers of their own ideological possibilities for their lives. This chapter takes up Bakhtin's iteration of the chronotope as a heuristic for considering how data from this connective ethnographic study of three teenagers' mobile phone-based composing practices demonstrate ideological possibilities. Specifically, in this chapter I examine how the participants actively produced time-space in their composing practices as evidenced by the nature of the practices and the spatiotemporal markers (both visual and discursive) in the resulting texts.

Keywords: Bakhtin, time-space, chronotope, selfie, mobile phone-based literacy practice, mobile phones, ideological becoming

This chapter explores mobile phone-based composing through the lens of Bakhtin's chronotope. Bakhtin (1981) borrowed from the theory of relativity the term chronotope (literally "time-space") to emphasize the inextricable nature of these two concepts. Rather than viewing time and place as the backdrop for interaction and communication among characters, he saw them as central and as indicative of an author's worldview. This perspective is useful in that it can help to challenge the apparent "naturalness" of literacy education as it currently exists in regard to time, space, and opportunity for learning in order to address associated pedagogical challenges.

For Bakhtin, the novel (and other representations of events and lifeworlds) invokes time and space and particularly how people live in time and space. He was interested in examining the ways in which the constructions of time and space by authors "index different ideological possibilities for characters and readers" (Sheehy & Leander, 2004, p. 6). Likewise, a chronotopic exploration of mobile phone-based composing practices can make visible the ideological possibilities that the participants hold for themselves as evidenced by their own constructions of time and space. In this chapter, I take up Bakhtin's iteration of the chronotope as a heuristic for considering how data from this connective ethnographic study of three teenagers' mobile phone-based composing practices demonstrate ideological possibilities. Specifically, in this chapter I examine how the participants actively produced time-space in their composing practices as evidenced by the nature of the practices and the spatiotemporal markers (both visual and discursive) in the resulting texts.

Chronotopes and Schooled Literacy

Time in the typical literacy classroom in the United States is carefully parceled and monitored. On a day-to-day basis in the classroom, "blocks" are uniformly scheduled and bells section off the beginning and end of each period. Schooling is based on "grade levels" and in the space of one year students are expected to advance from one level to another, where they study a different variation of language arts-related subject matter such as American literature, world literature, European literature, grammar, or composition. Terms related to schooling such as "graduation" which suggests a temporal division on a particular timescale reflects the importance of time within spaced for schooled literacy. Prescribed use of time is reinforced by the discourses of the "lesson plan" genre which is often organized around blocks of time as well as government and commercial curriculum materials organized by blocks of time.

Time is also a focal point in the literacy classroom in terms of the speed at which children should learn. The curriculum guides and lists of benchmarks and their accompanying tests through which learning is accomplished contain discourses that come from cognitive psychology and tell teachers and parents at what age students are capable gaining particular skills and knowledge and when (which we see in the convention of "grade levels"). If students do not demonstrate competence they are deemed "slow" in need of "remediation" terms that evoke a sense of time. Likewise, some learners may need "acceleration."

Literacy education is organized around specific conceptions of space. Literacy education is thought of as occurring at particular locations. The classroom is the ultimate space in which literacy learning occurs. However, literacy learning is longer thought of as limited to the insides of school buildings, and some literacy education researchers and practitioners recognize "out-of-school literacy practices". However, these are named with a term that suggests a physical space ("out-of-school") are still situated in physical locations like schools, homes, communities and workplaces. Popular research paradigms and design models in literacy education research are place-based as well, such as ethnography wherein the researcher goes somewhere physically to observe.

Literacy education is thought of as accomplished by specific and exacting uses of time, space, and materials. Learning is not thought to happen spontaneously in the dominant conception of literacy education, it is systematic. Personal literacy practices, those that occur among the family, or among peer groups or any combination or hybrid are not valued in the same way that literacy practices that are found in curriculum guides are valued, literacy practices for which time and space are rigidly controlled.

I contend that an analysis of mobile phone-based literacy practices must consider these practices within the framework of the chronotope in order to make apparent the ways that the organization of time and space in schooled spaces for literacy come to bear on the unsanctioned literacy practices of youth. In addition, an analysis of the interface between time and space with regard to these practices can make clear the ways that youth create or represent conceptions of time and space within their composing practices and the resultant texts they produce through those practices.

Mobility and Chronotope

While many scholars have focused on the temporal associations of chronotope (e.g., Bloome, Beierle, Grigorenko, & Goldman, 2009; Compton-Lilly, 2013), I seek to first examine the implications of place. Given that the participants in this study used mobile cellphones to compose, they created texts across the spaces they inhabited in their lifeworlds. Time and space were thus constituted in ways unique to composing by way of the mobile phone.

In some cases, the mobility of the participants' phones meant that they pushed back on dominant chronotopes. For example, in one instance, as I observed the participants' ninth grade English class taught by Ms. Harrison, I saw the participants interrupt the progression of the lesson to create their own learning paths. During a ninth grade English language arts lesson on the conventions of poetry, Ms. Harrison projected a question on the board about imagery in the particular poem the class was studying, Maya Angelou's *Phenomenal Woman*. Rather than rely only on the language of the poem to decipher the poem's meaning, as their teacher urged, the students in the classroom pulled out their iPads to Google-search photos of the poet to enhance their understanding of the poetic text:

> (*On the board is written: What images does the poet use? Why are these images important to the piece? Ms. Harrison directs the students' attention to these questions*).
>
> John: She said she confident; she said she ain't cute. (John cites a line in the poem about her hips).
> A student takes out her iPad and Googles Maya Angelou, then exclaims, "She not cute! She old!"
> Ms. Harrison: (Laughing) She wasn't always old!
> Student (to student with iPad out): Google a young picture.

In using Google and entering the online space during class to answer the teacher's question, they interrupted the chronotope of the lesson where communication was supposed to happen within a particular space (the physical classroom) and within particular timescales for the ideologically-imbued purposes of the teacher and the institution (which came from Ms. Harrison's lesson plans which were based on the state-mandated curriculum).

As was explored in chapters three, five, and six, the participants in this study represented their movement across spaces and their cellphone photography often captured their bodies in movement across spaces, as in Figure 5.2, posted by Eli, with the caption, "Walkin round wishin I hada female 2 chill wit haha :) ;)": In this example, time is represented as the present moment: "Walkin round wishin" is not "Walked around, wishin" or "Was walking around wishin." For Bakhtin (1981), time is never the catalyst for the action of the story; in the "adventure time" (the term he uses to describe time in Romantic stories), time is represented as insignificant, as merely "a logic of random disjunctions" (p. 90). The space is "interchangeable" (Bakhtin, 1981) with any other space, as it is in Eli's "adventure time." Like the adventure time of Romantic heroes, Eli seems to expect that events will happen to interrupt the status quo rather than taking initiative to enact change. These putative events are the temporal part of the chronotope. Eli represents time and space in his selfies in a very similar way, in a way that suggests his position in historical time and in the sociocultural space.

Selfies are ways of narrativizing social worlds and social worlds, are, of course, situated within time and space. In contrast to photography of the past, which was often about capturing what was, the self-taken self-portrait is about identity formation; as youth post selfies to social networks, they garner feedback that informs their burgeoning sense of who they are (Lenhart & Madden, 2007). The selfie sharing practice thus fosters the process of becoming. On social networks, "followers" can post feedback in response to postings through likes or comments and in this sense, the composing practice reflects a view of the self as object and self as malleable. The participants in this study, and in particular John and Eli for whom selfies were common practice, seemed to understand that the self should be informed by the social reception made evident through comments and likes. (In a similar way, Aaliyah used language and memes to communicate her identity and garnered feedback that informed her identity formation through likes, comments, and retweets from her online social connections).

The practice of deleting postings as a means of composing has chronotopic implications. When posted selfies garnered only a few likes, the participants would often delete them. The customary temporal relationship of the text and the production of the text was thus overturned as audience reception was folded into the composing process. Therefore, in either case, audience feedback happening on particular timescales informed the participants' ideological possibilities for themselves.

Other forms of smartphone photography engaged by the participants represented a means of creating a text in reflection on the past. For example, Eli posted a cellphone image of a framed photograph of his family before his parents divorced with the caption: "Lookin at diz picture makin me feel sum typa way." In other words, looking at the picture has him feeling emotional. He communicates his cognitive, social, and emotional stance in this posting and he demonstrates markers of his historical positioning as well. He evokes another configuration of time and space through this image. In doing so time "thickens, takes on flesh, becomes artistically visible; likewise, space becomes charged and responsive to the movements of time, plot and history" (Bakhtin, 1981, p. 84). In other words, the chronotope is at the center of his image in this example and central to his beliefs about the nature of his world, as well.

Figure 9.9: Reflecting on the Past.
Source: Author.

Curation and Chronotopes

As described in Chapter 7, John described using a feature called "favoriting" as a means of reflection. Favoriting is a practice wherein the participants read a text and click on it, thus grabbing it and adding it to their social media spaces, creating a new curated text. John often use favoriting to save tweets to an archive of sorts:

> When you favorite something…you can go back. It goes into, like your little, like a little secret box like on the internet or whatever. Just that only you can see, so if you wanna go look at it again. Or if you just goin', scrollin down through it, havin' like, a little memory time. That's why I favorite stuff. To see it again.

His use of favoriting afforded a means of digital memory that could be recalled and reflected upon, and that could persist into the future. John had a sense of the fact that these digital texts were chronotopic, evoking time and spaces past. For John "space becomes charged and responsive to the movements of time and history" (Bakhtin, 1981, p. 84) through his practice of favoriting and re-reading favorited texts.

Likewise, Aaliyah, a heavy curator of online content, selected, grabbed, and displayed content that represented ideas she agreed with or otherwise appreciated through retweeting on Twitter. Her retweets indexed her beliefs and worldview. Sometimes she would retweet a tweet wholecloth, and other times she would add her own commentary. In either case, Aaliyah reconfigured the chronotope by taking a previously published text and republishing it within a new timescale and context. She took text that was inert and brought it into new chronotopes imbuing them with her own ideological meanings, for her own purposes.

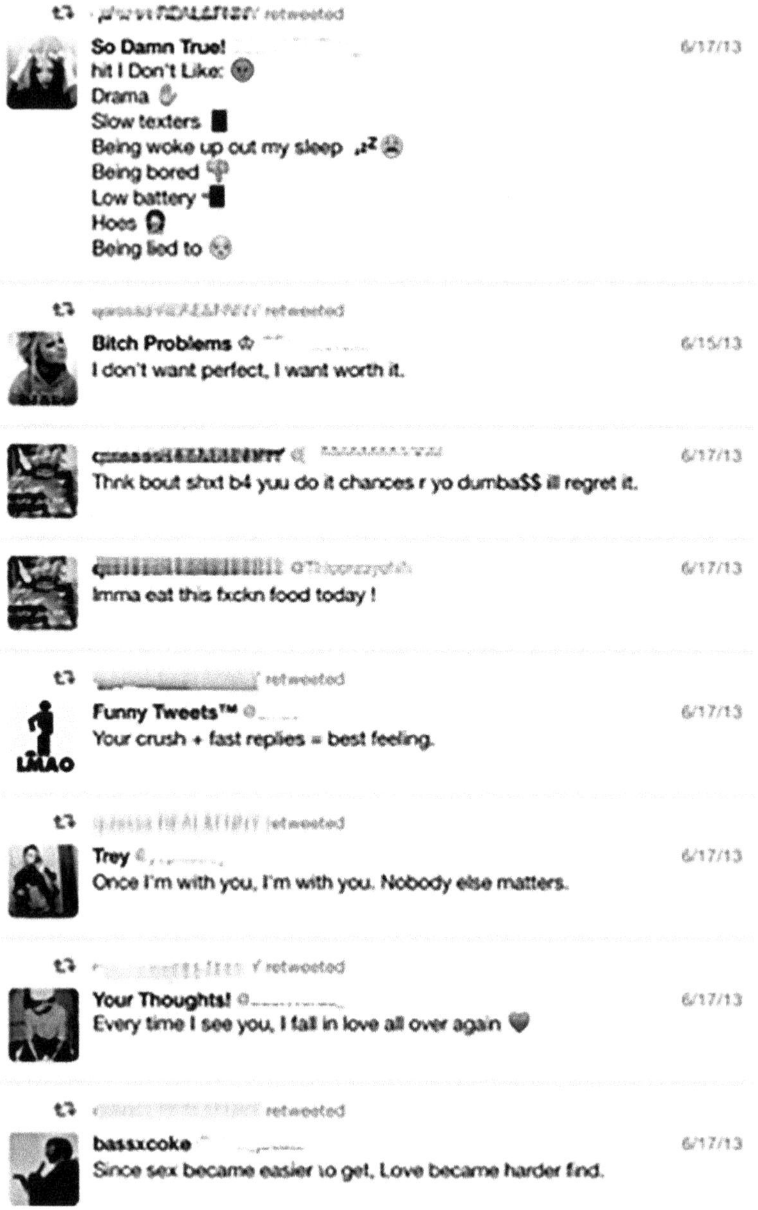

Figure 9.10: A Screenshot from Aaliyah's Twitter Feed with Mostly Curated Content. Source: Author.

Conclusions

For Bakhtin the hero was "one who is evolving and developing, a person who learns from life" (1981, p. 10). The participants were both authors and heroes as they narrativized their lives using methods of cellphone composing including selfies, mobile image sharing, favoriting, and curation. Through their practices they represent spatial and temporal formations that facilitate an understanding of how they see themselves and their relationship to their sociocultural world. The spatiotemporal configurations they deployed in their composing practices show their awareness of time-space conditions and the related implications for their identities and sociocultural positioning.

There is no grand, all-inclusive, coherent "genre" when it comes to mobile phone-based composing on any platform. In fact, users, as the heroes of sorts of the narrative inherent to lifestreams on Snapchat, tweet streams on Twitter, or profiles on Instagram, express time-space in different ways reflective of how they make sense of the world and their place in it. Time and space as they are reflected in the composing practices of youth reveal fundamental truths about learning and social development.

Furthermore, the participants did not always work within established chronotopes as they composed with and through their mobile phones. In fact, they often deterritorialized the spaces they inhabited online, disrupting the imposed order to invent new ways of using the media and creating their own conventions. The participants navigated both spaces and timescales; they often moved in "multiple, unpredictable directions of imagination, creation, and critique" (Leander, 2010, p. 206). The participants' practices unsettled predictable chronologies as they created texts with cellphones wherein time and space are evoked and indicate various ideological possibilities.

Sharing life experiences including past experiences, ongoing activity, or hopes and dreams for the future communicates ways of being in the world but it also gives shape and direction to composers' movements through time-space. Chronotopic analysis can help those interested in literacy development with and through mobile phones to see where school-constructed and self-constructed chronotopes are in conflict, which can allow for more agency for youth literacy and identity development. Future research in this area might focus on youth awareness of time-space in their composing practice in order to help them advance perceptions of their place in the world and their own development. Furthermore, rethinking chronotopes of schooled literacy learning can help us rethink or redesign conditions for learning in ways that can facilitate student understandings of their relationships to their sociocultural worlds.

References

Bakhtin, M. (1981). *The dialogic imagination*. Austin, TX: University of Texas Press.

Bloome, D., Beierle, M., Grigorenko, M., & Goldman, S. (2009). Learning over time: Uses of intercontextuality, collective memories, and classroom chronotopes in the construction of learning opportunities in a ninth-grade language arts classroom. *Language and Education, 23*(4), 313–334.

Compton-Lilly, C. (2013). The temporal expectations of schooling and literacy learning Jermaine's story. *Journal of Adolescent & Adult Literacy*, 56(5), 400–408.

Leander, K. (2010). Afterword. In Alvermann, D. (Ed.), *Adolescents' online literacies: Connecting classrooms, digital media, and popular culture* (pp. 114–129). New York, NY: Peter Lang.

Lenhart, A., & Madden, M. (2007, April 18). *Teens, privacy, and online social networks*. Retrieved from http://www.pewinternet.org/files/old-media/Files/Reports/2007/PIP_Teens_Privacy_SNS_Report_Final.pdf.pdf

Sheehy, M., & Leander, K. (2004). Introduction. In K. Leander & M. Sheehy (Eds.), *Spatializing literacy research and practice*. New York, NY: Peter Lang.

· 1 0 ·

PROPOSING AN EXPANDED UNDERSTANDING OF COMPOSING

Abstract

This chapter brings together the major ideas raised in the previous chapters and discusses their implications for literacy learning. Given this major shift toward social, collaborative, and dialogic composing amid a dominant paradigm of literacy that divorces composers and texts from their networked contexts, this chapter explores how those concerned with youth literacy fruitfully build on the strengths and repertories that youth bring into spaces for schooled literacy including the consideration of feedback, multimodality, and mobility. While schooled literacy has recognized that composing involves the integration of discourses (for example, persuasive writing as it is often framed by official curriculum, is highly addressive, and research writing is explicitly multivocal) composing is still framed as an individual accomplishment. This chapter argues that capitalizing on the tendency of youth to make meaning intertextually and dialogically would benefit youth in the pursuit of school-sanctioned literacy learning and presents directions for future research.

Keywords: Dialogism, composing, feedback, multimodality, mobility, youth repertories, mobile phone, literacy

In undertaking this research, I sought to describe the digital literacy practices of youth who were engaging not in niche, "exceptional," or fringe practices, but in more common composing practices encouraged by the mobile phone, taken up almost universally by youth today. Initially, I was aware that there was very little research that explored online, digital composing practices with mobile phones and I wanted to investigate the ways that youth were communicating with mobile devices because of the prevalence of such tools and the popularity of their concomitant supported practices. I sought to contribute new knowledge about such emerging practices, which are markedly different from the "new literacies" practices that have warranted study in recent years. For example, the ways in which youth think about and approach the act of composing using the phone with its particular affordances and constraints is different from computer-based digital media with less cumbersome keyboards as well as mouse access with all the convenience it affords for composing digitally. Perhaps most significantly, this study of youth composing with mobile phones showed that the traditional idea of writing process is challenged, curation (collecting and displaying digital texts) is a new form of authorship, and that youth rely on the voices of others as a central means of communicating with texts. The phone-based composing practices of the focal participants in this research were characterized by the rapidity of the composing process, the centrality of photography, the important role of audience feedback, and digital curation. In this closing discussion, I draw upon my dialogic framework to describe these findings and demonstrate their implications for literacy research and practice.

Composing Practice in Dialogue with Spaces

The youth's composing practices were in dialogue with the spaces, both digital and physical, which they inhabited, as they used their phones' cameras to capture objects and other material aspects of physical spaces to create texts or relied on existing voices and content in digital spaces to compose. In this way, rather than springing forth anew, their practices were thus reactive to the texts and the material aspects of the spaces both digital and physical that they inhabited. Instead of creating planned texts from scratch, they often created texts in order to respond to the world around them or responded to existing texts through multimodal means.

Not only did they respond to physical and material features in their composing spaces both digital and physical, the participants also responded to those with whom they interacted in digital spaces. For example, they revised their texts in response to feedback they received on those texts as they engaged in digital spaces. In fact, consideration of audience response or social reception was central to the youth's composing practice. Throughout this research, I noted the ways in which Bakhtin's dialogic model, posing that reception and production are reciprocal and mutually constitutive in that every utterance responds to or anticipates a response, was present in the data set. The youth used their mobile phones to write for, with, and through imagined and real audiences. In this way, in a very real sense, audiences co-authored texts.

Along those same lines, the participants learned through such dialogic interaction with the social world how to communicate in a way that would resonate or support understanding either through direct linguistic feedback or through the act of one passing on or favoriting a posting (Bakhtin, 1982, 1986). They learned how to interact, what would be understood, and what was socially appropriate though interaction, dialogue, and through reading and observing the replication of texts and conventions in digital spaces which offered cues about how to appropriately interact in these spaces. Through observation and reading spaces as well, they learned how to engage in digital spaces. They learned not only about the conventions of those spaces, but about how to make themselves known and communicate in ways that would allow them to connect with others.

Composing Practices in Dialogue with Other Texts

The data collected over the course of this research indicate that because of the major communicative affordances of the phone, including the camera, the cumbrous digital keyboard, and the architecture of the apps used by the participants such as the retweet and favoriting features, the participants relied heavily on digital curation as a means of composing. It offered less friction and was a means of communicating powerfully in nearly an instant. Digital curation involves selecting, compiling, and displaying existing digital content rather than creating texts anew. In other words, typing long texts or creating elaborate multimodal texts with a mobile phone is difficult, but mobile phone hardware and software include a multitude of different means for communicating with the press of a single button or through tapping a screen once or twice, and for this reason, it was a popular practice with the participants.

Digital curation as a means of composing may look lazy or like an alternative to true composing. However, while the participants' digital curation practice did indeed rely heavily on the texts and voices of others, it represents a kind of complex heteroglossic practice and is a scaffolded way for youth to begin to learn to compose online. And for those who are limited to the mobile phone as a tool for composing, it shows real resilience in that even without complex or expensive hardware or software, youth are engaging in so much rich communicative practice through a variety of multimodal means. Throughout this study, I was struck by resourcefulness inherent to this practice of using the mobile phone with its limited support for complex composing to participate in Web spaces in an increasingly digital world.

Just as curation as a composing practice has been much maligned, the agency of youth remixers has often been overlooked. Black (2008) explored this idea in depth in her work on fan fiction, digital composition that drew from existing texts instead of springing forth anew (which Bakhtin holds as impossibility anyway). "Far from being 'mindless consumers' and reproducers of existing media," Black asserts, "they actively engage with, rework, and appropriate ideological messages and materials of the original text" (p. xiii). While curation has not yet been accepted at scale as a legitimate literate practice, this study shows that it, at the very least, deserves further investigation. Further, by disassembling and reassembling culture through the process of curation, the participants demonstrated rich learning around the process of signification within multimodal texts and demonstrated addressivity (Bakhtin, 1982) by calling upon resources that carried relevance to the current moment and held particular meanings within an audience. Dissembling, rearranging, and otherwise tinkering with language, modes, and texts demonstrated their deep understanding of culture and expression (Jenkins, 2006a; Potter, 2012).

In addition to engaging in ongoing conversations (Bakhtin, 1982) online by creating, responding to, and circulating existing content, the youth in this study created their own original practices and conventions. Whereas a technological determinist view would hold that youth are constrained by the existing architecture in the digital environments in which they compose, the data collected in this research showed the ways youth took up the tools and affordances appropriate and suited to their aims. Not only did they create their own texts, but they created and shaped their means of text production through highly inventive social practice with digital tools.

Of course, youth who are primarily using the voices of others to express themselves through digital curation are limited in some ways by the available voices. In the practice of curation, youth engaging heavily in digital curation because of the lack of access to tools with more convenient keyboards, for example, are limited in this way. In the practice of curation, it can be argued, youth's perspectives, views, and unique voices are in a subordinate position to existing cultural artifacts or resonant voices in the digital spaces they inhabit. This suggests that there is a need to teach critical digital literacy wherein youth are encouraged to interrogate the discourses and voices that permeate the texts they draw from in their own composing practices.

Unfortunately, as of now, while it is true that composing using existing texts can act as a scaffold to production from the ground up, and is not a substitute for but exists in addition to true rendering, it is rendering, not curating texts, that is privileged and rewarded in schooled literacy contexts. As I observed over the course of this study, the kinds of texts the participants and their classmates were asked to produce in their English class were solo-authored, and they were assessed individually. In school, appropriation without attribution like much of the remixing that goes on in the online space constitutes plagiarism or cheating. Online, however, appropriating the ideas and products of others is prevalent and even encouraged. In fact, on the web, this kind of appropriation is a form of authorship in and of itself, and is seen as a legitimate way of engaging in this space. It is true cultural production, whether or not it would be acceptable and rewarded practice in schooled literacy spaces.

Proposing an Expanded Understanding of Composing

Schooled literacy has recognized that composing involves the integration of discourses (Bakhtin, 1982). Persuasive writing as it is often framed by official curriculum, is highly addressive, and research writing is explicitly multivocal. Yet, schools continue to teach, for the most part, as if literacy is an individual accomplishment and that literate practice is the representation of a cognitive, rather than a social, process. Whereas co-opting the voices of others is sanctioned in the digital space, where digital curation is popular practice, in school, it can be a punishable offense. Classroom practice that includes the collection, critical reading of, response to, and assemblage of diverse voices can build on the proficiencies that youth are building through online curation practice. Teens engage in phone-based literacy practices like microblogging and social photo sharing because they are afforded feedback. As Hicks (2013) points out,

> The art and skill of composing something as seemingly simple as a tweet is more than simply typing a limited number of characters into a small box … as students compose texts, social media help them develop ideas, give and receive feedback, and ultimately share their work with a worldwide audience. (pp. 142–146)

Hicks suggests that teachers "have … students create significant texts … and share links to their work, inviting feedback from their classmates and perhaps beyond the classroom as well." (p. 146). Bringing social media into the classroom, even just through a school-contained platform like Edmodo (a free social networking platform for students that can have strict privacy settings, and which can be loaded as a mobile app onto a smartphone) can afford students the kind of powerful feedback that they search out through their use of social media. And in learning to use feedback and others' voices in the composing process, it would benefit students to have the opportunity to learn how to read and think critically not only about media messages but about how digital tools work and are enmeshed in larger structures of power.

In addition, to examining the dialogism inherent to digital practices and texts, the mobility of digital literacy practices with and through mobile phones offers a rich site of exploration for schooled literacy learning. Although technologies are being taken up with increasing frequency in schooled spaces for literacy learning, they are often taken up in ways that mirror print literacies. It is very rare, for example, for technology to "go mobile" in the classroom. During a technology-infused lesson, it is still commonplace to see student desks in rows, and students working independently on devices, even mobile devices, while seated alone at their desk or in a computer lab. However, as the data in this study showed, literacy happened in and across concentric spaces, and was uncontrolled and spontaneous. New literacies are not static or moored.

The multimodality of mobile phone-based composing practices is also important to consider in thinking about reconceptualizing composing practice. Likewise, linguistic text is still what is privileged in the schooled literacy context. Although providing a variety of modes and a wider semiotic toolkit would mean that youth who struggle with schooled literacy would have a richer toolkit for meaning making and communicating in the classroom, it has proved problematic for a variety of reasons including curricular constraints, a lack of appropriate tools, and a lack of teacher training, to bring multiple modalities and multimedia into schooled spaces for learning. Gardner and Davis (2013) acknowledge the "creative potentials of apps and other forms of digital media" and lament the narrowness of what counts in school in light of the "increased sophistication of what youth can create, and the wider variety of creative opportunities now open to young people" (p. 149). It would benefit youth to be allowed to communicate in means that go beyond print texts, including the visual. Affording students as broad of a semiotic toolkit as possible will allow more students to participate as literate beings. For example, photography can be an entry point for teaching school-sanctioned writing like narrative, expository, and even persuasive by having students tell stories through original photos or by having students write the story of a photograph using linguistic text. Allowing for the integration of media forms across a variety of spaces in the teaching of formal schooled writing would grant more pathways for meaning making in the schooled space, and perhaps through some of the increased federal attention to the need for STEM teacher training and funding for tools and infrastructure will help make this a reality.

Not only are some media and modes pushed to the margins when it comes to schooled literacy, but some communicative practices, particularly those encouraged by the affordances of the mobile phone, are still seen as frivolous and even silly. This is often the case with any practice that is popular with young people or emblematic of youth culture, likely because adults lack an understanding of the social meanings of those practices. Ito and colleagues (2010) suggest in their exploration of the overlap of "'serious' online information and culture" and "social and recreational activities online" that there is a necessary "cultural shift" that needs to happen within formal institutions for learning that would allow for an array of opportunities for learning denied by a view of youth digital practice as merely frivolous (p. 347). While smartphone-friendly classrooms are not yet universally available to students, there are a number of ways that educators might build the on phone-based literacy practices of youth, rather than having students check such proficiencies at the door.

The mobile phone-based communicative practices explored in this book are also representative of the kind of deep literacy learning that can happen just through engagement in a space and a practice, even without direct instruction around how to communicate. The youth in this study demonstrated tacit social learning of communicative conventions and came to understand what constituted effective communication in online spaces through interaction, perhaps the provision of more opportunities for composing for real audiences that afford timely feedback would serve students who compose digitally with fervor yet struggle with schooled literacy. This is what Ito et al. (2010) characterized as "highly active forms of learning" (p. 291). Capitalizing on the tendency of youth to make meaning intertextually and dialogically through authentic engagement in digital spaces with and through their mobile phones would undoubtedly benefit youth in the pursuit of school-sanctioned literacy learning.

Mobile Phones and Classroom Practice

According to Purcell, Heaps, Buchanan, and Friedrich (2013), 73% of middle and high school classroom teachers integrate cell phones into classroom instruction. While in the past, schools have policed the use of mobile phones, doing everything from placing restrictions on their use to outright banning phones, the perceptions around the place of phones in the classroom has changed. Mobile phones are still often seen as a distraction or at worst tools that facilitate bullying or cheating, but increasingly schools and teachers are beginning to recognize the learning potential of phones. Until the discourse around mobile phones in the classroom moves away from classroom management to focus on learning, the true potential of phones for learning will not be realized.

While open mobile phone policies as of yet rare, the stigma around mobile phones has faded somewhat. Just a few years ago, phones were almost universally banned from classrooms, but as of the time of this writing, 70% of school districts have lifted phones bans. Even New York City, the largest school district and the district that grabbed headlines because of the issues caused by banning phones including the emergence of mobile phone "lockers" which became a huge industry in New York, lifted their mobile phone ban. Yet, there is no consensus among educators around how best to manage mobile phones in classrooms, and the onus is usually on individual teachers to decide how and when to allow phones to emerge from student backpacks and pockets. It may be this inconsistency that causes so much confusion around how and even if mobile phones have a place in schooled spaces for literacy.

Further Research

Recent research on youth digital literacy practice has confirmed that teens are doing more than "hanging out" and "messing around" (Ito et al., 2010) using their digital tools and that widespread fears about bullying, loss of privacy, and other online dangers are misconceptions that obscure the agency, important identity work, and critical thinking teens mobilize in their digital practices (Boyd, 2014). Further research on the implications of the practices described in this article with regard to the body of work on the digital divide is needed given the near ubiquity of these practices among youth and widespread mobile adoption as the primary form of Internet access. And as more schools are moving to Bring Your Own Device models (St. George, 2014) and lifting classroom bans on cellphones (Rich & Taylor, 2014), teachers need ways to generatively marshal the affordances of cellphones for formal literacy learning (without colonizing or appropriating youth practices for the teacher's goals). The field of literacy research will need to address these issues given this major shift toward social, collaborative, and dialogic composing amid a dominant paradigm of literacy that so often divorces composers and texts from their networked contexts.

To carry out such research, though, new frameworks are first needed to allow for the recognition and representation of the richness of these practices. The Bakhtinian framework taken up in this study was useful in framing the multivoiced, intertextual, and dialogic nature of the youth's literacy practices. However, there is so much social practice outside of that which deals with language and voices within these phone-based composing practices for which a Bakhtinian framework cannot account. For example, beyond the communicative elements explored in this volume, curation evokes categorization and hierarchical thinking. In addition, the mobility of the youth practice was not fully compatible with the dialogic framing employed in this study. Bakhtin's focus on spoken and written language is not fully compatible with a discussion of multimodal texts. We need new frameworks to describe these transspatial, multimodal, socially rich new practices happening with and through mobile technologies if we are to push the field forward in an age where so much rich literacy learning is happening with and through mobile phones.

The New Literacy Studies, which challenged the view of literacy as the mastery of decontextualized skills, in many ways transformed what teachers viewed as legitimate literate practice in the lives of students. The data collected in this study suggest that literacy researchers and practitioners, in an age of almost universal smartphone use, need an even broader view of literacy practice than that the New Literacy Studies affords. The New Literacy Studies are rooted in the idea of design, proposing that youth are making choices about how to design texts, putting planning and forethought into their digital compositions. My findings indicate that because of the ubiquity of the phone, made possible by its mobile nature (always on participants' person), much of participants' literacy engagement (such as selfies, retweets, and conversations with friends on photos) was spontaneous, in the moment, reactive, and emergent rather than focused on an end product (cf. Leander & Boldt, 2013). The New London Group's (1996) pedagogy of multiliteracies considered that emerging technologies contributed to the appearance of a variety of new text forms and new practices, but my findings indicate that the theory is too narrow, pivoting on the idea of youth as rational designers of texts. The youth in this study often created texts as a means of emotional release, using composing tools like cameras with which they could easily, quickly, and fluidly communicate powerfully through a visual image; their composing was often done without much observable forethought. Instead, they oftentimes contemplated and revised texts after they posted them to their web spaces.

Considering the popularity of the practice of digital curation, I argue that youth who engage in this practice are being socialized into a space with a philosophy of language that is in contention with that of schooled literacy. Whereas the participants in this study borrowed and remixed texts without attribution as normal practice, in school such a practice would be considered plagiarism. Whereas coopting the voices of others is a legitimized way of participating in the digital space, where digital curation is popular practice, in school, taking up the voices of others to express one's ideas is seen as plagiarism or academic dishonesty. Youth would benefit from opportunities to critically analyze the perspectives of other thinkers and to speak through the voices of others as they work to assemble their emerging identities while learning about academic citation conventions and mobilizing their voices for participation in society.

Furthermore, critical analysis of the voices present in youth lifeworlds including digital spaces is, as my findings show, critical now. As is evident in the case of John, for example, he often disrupted the circulation of harmful ideologies through his productive subversion (reclaiming the term "nappy head" as a means of pride and by creating a gathering space around this pride), he also at times did not seem to recognize the harmful ideologies he circulated (e.g., "hoes"). Helping students to recognize the circulation of discourses should be part of critical literacy practice for the 21st century. Particularly since contexts for composing often have a more permanent nature than many youth realize, such critical analysis is crucial since it can have an impact on youth's social futures as texts persist in the digital space, and even, once deleted, live on in servers often owned by private industry. Critical digital literacies can offer youth more power and agency in light of this new reality around spaces for composing.

Finally, the obfuscation of advertising in the spaces where participants interacted, including school, and the texts they read (e.g., product placement in music videos) manifested in participants' discourse. Participants subtly and overtly advertised brands, companies, and even their own friends in their online writing, and took up discourses of commercialism. Critical literacy practice for the 21st century must teach young people to recognize advertising as a practice that positions them as consumers. While youth often use identification with brands as a means of identity construction and communication, they may not always recognize the implications of postmodern in an increasingly advertised-saturated world.

Lankshear and Knobel (2006) differentiated between the new "technical stuff" (hardware and software) and "new ethos" stuff (p. 93), which they defined as "the emergence of a distinctly contemporary mindset" (p. 73) arising in conjunction with the social use of new technologies. In doing so, they distinguish between the technologies themselves and how people take use them. Coiro, Knobel, Lankshear, and Leu (2008) made a similar distinction between the material aspects of technology and the "important new strategies and dispositions" (p. 10) they inspire (p. 1161). Coiro et al. (2008) argued for the existence of "changes in social and cultural ways of doing things, ways of being, ways of viewing the world (world views), and so on" (p. 7). The Internet is a "literacy issue" (Leu, O'Byrne, Zawilinski, McVerry, & Everett-Cacopardo, 2009), and the new ethos centers mainly on the composition, circulation, and consumption of texts. Approaches to composition with new tools in new spaces demand the attention of literacy researchers. Yet, little extant research describes teenagers' perspectives and practices in the creation and dissemination of online multimodal texts and specifically their dialogic practice in networked, multimodal, and potentially interactive and global new spaces. Researchers of youth digital composing have been preoccupied with describing and cataloging the output of young people through digital venues while grieving lack of engagement with schooled literacy. This is a useful starting place, but the associated worldviews, ways of being, and social and cultural ways of acting are also important to aid in understanding youth media composers as learners.

Because the demands of online digital writing contexts on young composers are so different from those of the print-based writing contexts they typically encounter in school, especially those engaged in with and through mobile phones, it seems regretful that most schools continue to teach predominantly in the framework of the ethos of print literacies (Gee, 2004; Lam,

2006; Sefton-Green, 2006), even though most contemporary communication is mediated, taking place over email, on message boards, text messages over hand-held devices, and through social networking and other websites. Certainly, students like the participants in this study are not positioned in a way that allows them to be successful when the literacy curriculum ignores their primary means and modes of communication and fails to meet them where they are, honoring the strengths that could help them be successful in schooled literacy contexts. Most students currently in primary and secondary schools communicate primarily using mobile phones and experience a disconnection between these competencies and what is rewarded in the schooled literacy context. The privileging of written language deprives students of useful semiotic resources for meaning making. Therefore, without descriptions of the practices, perspectives, and approaches to composing of the digital multimodal composing practices of youth, students who are "multimodally literate" remain unrecognized. Siegel (2012) reminded that "multimodal transformations of school literacy are a matter of social justice" (p. 674) because the more tools that diverse students have in their communicative toolkits, the wider the circle of who can participate and enact literacies.

To address the disconnection between the new ethos and the narrow, government-mandated literacy curriculum, more studies that examine specific literacy practices and cultural models characteristic of the new ethos need to be conducted. Not only do people lack understanding about how youth negotiate the demands of new composing environments, but they also lack knowledge about how composers approach composition using resources other than language in new spaces. In addition to an expanded definition of literacy, a deepened understanding of adolescents as multimodal composers and how such an understanding might prove useful in designing contemporary curriculum and teaching is critical now, as verbocentric "assessment culture" (Kontovourki, 2012) and the ways youth communicate (as well as what is necessary to be literate in postmodern, digitally-saturated culture) are at odds.

In other words, the scarcity of attention to the new ethos of digital literacies, especially those encouraged by the affordances of the mobile phone given their popularity with youth, not only hinders the literacy learning of those students who approach literacy tasks from this perspective, but also ensures that students who do not have this mindset and the concomitant strategies and cultural models will be deprived of the chance to develop them. This is problematic given assertions that this new ethos will be needed to achieve success in an increasingly digital, networked, and globalized society and in their futures. Educators faced with the task of developing literacy must be able to respond to these shifts (Reinking, 1998), no matter how rapidly things change within the digital landscape.

Conclusion

I first began to think about the ideas that later coalesced into this book when I met Anthony. I met Anthony when he was a student in the tenth grade college-preparatory English class I taught at a public high school in metro-Atlanta, Georgia that, due to years of underachievement by government standards, was in danger of losing federal funding. He became part of the group of youth I called my "regulars," those who socialized in my classroom at lunch, between classes, after school, and even before if they could enter the building, whether or not I was teaching them that semester. Outgoing, loquacious, creative, and eager, Anthony did well in my class but, like most of the students at the high-poverty school, was not adept at passing the state-mandated standardized tests. I was troubled by his struggles with official English language arts curriculum materials and assessments, because I knew he was highly literate in many capacities, especially with mobile devices (his grandmother worked for a cell phone company), which he frequently illustrated to me in a friendly show-and-tell. He was also a heavy user of my classroom computers, because, as he told me, he did not have Internet access at home, and his grandmother's computer was "too slow for graphics."

One day after school was out, distracted from my grading by the sound of the furious tapping of keys, I became curious about what Anthony was doing on my classroom computer. He explained he was working on his Web log or blog, which he updated almost daily, and sometimes several times a day. The posts in a blog, which are time-stamped and displayed in reverse chronological order, are primarily text-based but are usually multimodal in ways that extend beyond the multimodality of print texts, comprised of additional digital modes such as still images (Lutkewitte, 2013), moving images (cf. Mills, 2011), hyperlinks (Jewitt, 2005), and more. Another characteristic of Anthony's blog was the commenting feature, which allowed readers to respond to his blog posts, engaging in a conversation. The blog roll, or list of interlinked blogs (those he read or actively commented on), is another dialogical feature of Anthony's blog in addition to commenting, allowing for conversation and community.

Reading Anthony's blog, I found diary-like entries about his day-to-day activities, musings and diatribes about friends and school, poetry and song lyrics, embedded video culled from video sharing sites like YouTube, images he drew or found online, and photographs he had taken with his cell-phone camera. The choice of design, typography, and layout worked to communicate, as well. He updated his blog, writing new posts or posting new content with great frequency, and, he told me, changed the layout and design at times to "match [his] mood."

Even ten years ago, in his online digital composing practice, Anthony was not atypical. According to the 2007 Pew Internet and American Life study, 28% of adolescents at that time had created their own online Web log, up from 19% in their 2004 report. The 2007 Pew study concluded that 64% of teens could be considered media creators, blogging, posting their artwork, music, videos, photography, or remixing other content they find online to create new original work, an increase compared to the 57% of teens found to be content creators in the 2004 study (Lenhart & Madden, 2005; Lenhart, Madden, Smith, & Macgill, 2007).

This is, in other words, not a new phenomenon. This tendency of youth to use technologies to compose dialogically and with multiple modalities across spaces exemplifies what Jenkins (1992, 2006a, 2006b) called "participatory culture," wherein people are actively creating media instead of just consuming it, a criticism of previously dominant broadcast media originally raised by Brecht (1931/1967) and Enzensberger (1970). This movement has also been likened to the do-it-yourself movement in art, music, and media (Guzzetti, Elliot, & Welsch, 2010; Knobel & Lankshear, 2010; McKay, 1998; Quart, 2013) originating in punk culture (Triggs, 2006) or perhaps even earlier with the existence of pirate radio (Walker, 2004). This is not new, and we cannot continue to ignore it.

In fact, during my time as a high school teacher, I noted many of my students actively disseminating their ideas and creative works via blogs, personal Web pages, social networking sites, fanfiction sites, file-sharing sites, and text messages. Their digital media content was often comprised of remixed or "mashed-up" content or content otherwise reactionary to mainstream culture, including school culture. Rather than passive consumption, participation, and responding to cultural messages had already become the new norm. And as hardware becomes has become progressively more affordable, technologies have continued to become more ubiquitous and less cumbersome, and accordingly, we are composing in and across spaces more than ever before.

Initially, I was most fascinated upon realizing that Anthony's blog, hosted on Google's popular blogging platform Blogger.com, could potentially find readership anywhere in the world where there was access to the Web. The blog posts he composed in my classroom in that Atlanta suburb could traverse not only geographic boundaries, but also cultural, temporal, and even linguistic boundaries. Sometimes referred to as "scalability," this capacity for computer-mediated communication to reach or become accessible to a larger audience than is possible with offline texts, is a possibility for Anthony's and others' blogs and other Web texts. This realization of the scaling potentialities of Internet publishing led me to wonder: How did Anthony approach the writing task when writing online for an invisible readership of potentially everyone and consequently no one in particular in an age of global flows (Appadurai, 1996)? Considering that his school-sanctioned writing was not considered strong by government standards, I wondered about the shortcomings of a system of schooling that would label as deficient a writer with a wide and active readership, who negotiated communicative conventions in a global space of blurred cultural and other boundaries.

Furthermore, semiotically, Anthony's blog text was quite sophisticated. He used hyperlinks in ways that suggested he understood the networked nature of his text. He integrated multiple modalities in a way that suggested he had awareness of the respective communicative powers of various modes and chose among them purposefully. He interacted with others in the space of his blog, which suggested he understood the spatial qualities of the text, such as the text as not just an artifact, but also as a meeting place where people can share ideas. Moreover, because he was interested in German culture, a love that began in his German classes at the high school, he integrated German cultural artifacts from around the Web, suggesting he understood the text to have global scalability. Through the case of Anthony, I first became aware of some of the early ways young people were using mobile phones to communicate beyond simple one-to-one messaging.

I tell this story not just to illuminate the genesis of this research and to more clearly link the research and theory in these pages to practice, but also to frame this work as "nothing new." Manovich (2001) has warned about declaring ways new media communicate "new," as there is little new about so-called new media, and consideration of new media must be carried out "in relation to a number of other areas of culture" such as other arts and media, the materiality of modern computers, today's visual culture, and how a culture encodes information: "information culture" (p. 13). Manovich's approach is in contrast to the dominant one in literacy education research, which often attempts to equate multimodal communication to a language, a perspective influenced by scholars such as Kress (2003, 2010) and Van Leeuwen (2005) who liken communication on new media to a structuralist system. New media are cultural objects (Manovich, 2001) and thus must be understood not in its discrete parts, but in their social use in an ecology that includes material and cultural aspects, including the "ways in which information is presented in different cultural sites" (Manovich, 2001, p. 13), or information culture. I see the mobile phone based literacy practices explored in these pages as squarely tied to a long tradition of the unsanctioned literacy practices of youth looking to learn about themselves and find a way of being in the world with and through others.

Yet in some ways, mobile phone-based practices are very new. Anthony's phone allowed data to travel in ways that facilitated interaction across broad spaces in real time. His phone also made it possible for him to produce spaces socially. One can now create a website and upload it to the Web, where it has the potential to become a site of interaction. Anthony, for example, used technology to create a space in which he was an adept, literate being. This space was separate from the school space defined by the material place of the school and the texts that were assigned to define him in that space. And as powerful technologies have become less cumbersome, they have gained increased mobility. The computing power of an iPhone, sitting innocuously in the pockets of millions of people, would have been unthinkable just 40 years ago when a computer would fill a room. Now, people communicate through bits and bytes as they perform their daily tasks, wherever they can carry their primary devices, which are smaller and more convenient. The mobility of such interfaces encourages constant connection and the potential for communication across physical spaces at any time.

Because the demands of phone-based digital writing contexts on young composers are in many ways so different from those of the print-based writing contexts they typically encounter in school, it is at best and oversight and at worst negligent that schools continue to teach through the framework of the ethos of print literacies (Gee, 2004; Lam, 2006; Sefton-Green, 2006). Certainly, students like Anthony or Eli or John or Aaliyah are not positioned to allow them to be successful when the literacy curriculum ignores their primary means and modes of communication and fails to meet them where they are, honoring the strengths that could help them be successful in schooled literacy contexts. Without rich descriptions of the practices, perspectives, and approaches to phone-based digital multimodal composing practices of youth, youth who are adept in this realm remain unrecognized.

Not only do we, for the most part, lack understanding about how people negotiate the demands of phone-based composing environments, but we also lack knowledge about how composers approach composition using many of the emerging resources other than language in new spaces. In addition to an expanded definition of literacy, a deepened understanding of students as multimodal composers stemming from platforms common to mobile phones and how such an understanding might prove useful in designing contemporary curriculum and teaching is critical now, as curriculum shrinks to accommodate a culture of testing and accountability (Kontovourki, 2012) and the ways youth commonly communicate (as well as what is necessary to be literate in postmodern, digitally-saturated culture) are at odds.

The purpose of the research in these pages was to describe the practices, perspectives, and approaches to composing of three teenagers with a history of school underachievement who composed in networked, multimodal, potentially interactive and global new media environments using a variety of modes with their mobile phones. In particular, the purpose of the study was to describe how their practices reflect and recirculate tacit, socially-shared approaches related to composing with mobile phones in and across networked spaces. Further, I sought to describe the social meanings they made of their literacy practices across spaces.

Through this work, I aim to contribute to a growing body of work foregrounding the movement of bodies in and through social space(s) (e.g., Leander & Boldt, 2013) in literacy practice. With increasing frequency, the literacy practices of youth involve digital and networked media, which are inherently social and often mobile. This study adds to a body of work that presents a spatial dimension of youth literacy practices, neglected when the focus is reduced, as it has been traditionally in the field of multiliteracy studies, to describing the visual grammar and parsing multimodal texts at the expense of examining how they use the creation of texts and the texts themselves to mediate their human experience. I aimed to present the practices related to composing with digital media that goes beyond describing a metalanguage to interpret what these literacy practices help youth to accomplish and to explore how this can inform a vision of responsive literacy education.

In my own classroom, I have asked students to use their phones to access educational applications, to tweet responses to readings and as a form of classroom discussion, and even to use their cell phone cameras to visually demonstrate and document learning. Of course, to harness cell phones in the classroom for learning, teachers have to tread a fine line, capitalizing on the capabilities of these tools without making students feel that youth practices are being coopted or usurped. Responsive teacher training must address ways of bringing mobile tools and multimodal texts into the classroom that build on the skills youth bring as frequent and dexterous digital composers rather than simply importing youth practices, fully intact, for the teacher's own goals. With the social turn in education, educators sought to recognize the complex social circumstances of students. Honoring students' funds of *digital* knowledge, however educators can best accomplish that task, is vital now.

References

Appadurai, A. (1996). *Modernity at large: Cultural dimensions of globalization*. Minneapolis, MN: University of Minnesota Press.

Bakhtin, M. (1982). *The dialogic imagination*. Austin, TX: University of Texas Press.

Bakhtin, M. (1986). *Speech genres and other late essays* (V. W. McGee, Trans.). Austin, TX: University of Texas Press.

Black, R. W. (2008). *Adolescents and online fan fiction*. New York, NY: Peter Lang.

boyd, d. (2014, March 21). Why snapchat is valuable: It's all about attention. [Web log]. Retrieved from http://www.zephoria.org/thoughts/archives/2014/03/21/snapchat-attention.html

Brecht, B. (1967). der dreigroschenprozeß [The Three penny Opera]. In B. Brecht (Ed.), *gesammelte werke* [Collected works] (Vol. 18, pp. 136–156). Frankfurt, Germany: Suhrkamp. (Original work published 1931)

Coiro, J., Knobel, M., Lankshear, C., & Leu, D. J. (2008). Central issues in new literacies and new literacies research. In J. Coiro, M. Knobel, C. Lankshear, & D. Leu (Eds.), *Handbook of research on new literacies* (pp. 1–21). Mahwah, NJ: Lawrence Erlbaum.

Enzensberger, H. M. (1970). Constituents of a theory of the media. *New Left Review*, 64, 13–36.

Gardner, H., & Davis, K. (2013). *The App Generation: How today's youth navigate identity, intimacy, and imagination in a digital world*. New Haven, CT: Yale University Press.

Gee, J. P. (2004). *Situated language and learning: A critique of traditional schooling*. New York, NY: Routledge.

Guzzetti, B., Elliot, K., & Welsch, D. (2010). *DIY media in the classroom: New literacies across content areas*. New York, NY: Teachers College Press.

Hicks, T. (2013). *Crafting digital writing: Composing texts across media and genres*. Portsmouth, NH: Heinemann.

Ito, M. (Ed.). (2010). *Hanging out, messing around, and geeking out: Kids living and learning with new media*. Cambridge, MA: MIT Press.

Jenkins, H. (1992). *Textual poachers: Television, fans, and participatory culture*. New York, NY: Routledge.

Jenkins, H. (2006a). *Convergence culture: Where old and new media collide*. New York, NY: New York University Press.

Jenkins, H. (2006b). *Fans, bloggers, and gamers: Exploring participatory culture*. New York, NY: New York University Press.

Jewitt, C. (2005). Multimodality, "reading," and "writing" for the 21st century. *Discourse: Studies in the Cultural Politics of Education, 26*, 315–331.

Knobel, M., & Lankshear, C. (2010). *DIY media: Creating, sharing, and learning with new technologies*. New York, NY: Peter Lang.

Kontovourki, S. (2012). Reading leveled books in assessment-saturated classrooms: A close examination of unmarked processes of assessment. *Reading Research Quarterly, 47*, 153–171. doi:10.1002/RRQ.014

Kress, G. (2003). *Literacy in the new media age*. London: Routledge.

Kress, G. (2010). *Multimodality: A social semiotic approach to contemporary communication*. London: Routledge.

Lam, W. S. E. (2006). Culture and learning in the context of globalization: Research directions. *Review of Research in Education, 30*, 213–237. doi:10.3102/0091732X030001213

Lankshear, C., & Knobel, M. (2006). *New literacies: Everyday practices and classroom learning* (2nd ed.). New York, NY: Open University Press.

Leander, K., & Boldt, G. (2013). Rereading "a pedagogy of multiliteracies": bodies, texts, and emergence. *Journal of Literacy Research, 45*, 22–46. doi:10.1177/1086296X12468587

Lenhart, A., & Madden, M. (2005, November 2). *Teen content creators and consumers*. Retrieved from http://www.pewinternet.org/~/media/Files/Reports/2005/PIP_Teens_Content_Creation.pdf.pdf

Lenhart, A., Madden, M., Smith, A., & Macgill, A. R. (2007). *Teens and social media*. Retrieved from http://www.pewinternet.org/Reports/2007/Teens-and-Social-Media.aspx

Leu, D. J., O'Byrne, W. I., Zawilinski, L., McVerry, J. G., & Everett-Cacopardo, H. (2009). Comments on Greehow, Robelia, and Hughes: Expanding the new literacies conversation. *Educational Researcher, 38*, 264–269. doi:10.3102/0013189X09336676

Lutkewitte, C. (2013). *Multimodal composition: A critical sourcebook*. Boston, MA: Bedford/St. Martin's.

Manovich, L. (2001). *The language of new media*. Cambridge, MA: MIT Press.

McKay, G. (1998). *DIY culture: Party & protest in nineties Britain*. London: Verso.

Mills, K. (2011). "Now I know their secrets": Kineikonic texts in the literacy classroom. *Australian Journal of Language and Literacy, 34*, 24–37.

New London Group. (1996). A pedagogy of multiliteracies: Designing social futures. *Harvard Educational Review, 66*, 60–92.

Potter, J. (2012). *Digital media and learner identity: The new curatorship*. New York, NY: Palgrave Macmillan.

Purcell, K., Heaps, A., Buchanan, J., & Friedrich, L. (2013). How teachers are using technology at their homes and in their classrooms. Washington, DC: Pew Research Center's Internet and American Life Project.

Quart, A. (2013). *Republic of outsiders: The power of amateurs, dreamers, and rebels*. New York, NY: New Press.

Reinking, D. (1998). Synthesizing technological transformations of literacy in a post-typographic world. In D. Reinking, M. C. McKenna, L. D. Labbo, & R. D. Kieffer (Eds.), *Handbook of literacy and technology: Transformations in a post-typographic world* (pp. xi–xxx). Mahwah, NJ: Erlbaum.

Rich, M., & Taylor, K. (2014, October 31). With school ban nearing end, New York City works on how and when to allow cellphones. *The New York Times*. Retrieved from http://www.nytimes.com/2014/11/01/nyregion/with-school-ban-nearing-end-new-york-city-works-on-how-and-when-to-allow-cellphones.html

Sefton-Green, J. (2006). Youth, technology, and media cultures. *Review of Research in Education*, *30*, 279–306.

Siegel, M. (2012). New times for multimodality? Confronting the accountability culture. *Journal of Adolescent & Adult Literacy*, 55, 671–681.

St. George, D. (2014, September 14). Schools move toward 'Bring Your Own Device' policies to boost student tech use. *The Washington Post*. Retrieved from http://www.washingtonpost.com/local/education/stem/schools-move-toward-bring-your-own-device-practices-to-boost-student-tech-use/2014/09/14/4d1e3232-393e-11e4-9c9f-ebb47272e40e_story.html

Triggs, T. (2006). Scissors and glue: Punk fanzines and the creation of a DIY aesthetic. *Journal of Design History*, *19*, 69–83. doi:10.1093/jdh/epk006

Van Leeuwen, T. (2005). *Introducing social semiotics*. London: Routledge.

Walker, J. (2004). *Rebels on the air: An alternative history of radio in America*. New York, NY: New York University Press.

BIOGRAPHY

Julie Warner is an independent researcher who explores the intersection of technologies and literacies. She is also a Presidential Management Fellow and Education Research Analyst in the Office of Planning, Evaluation, and Policy Development at the United States Department of Education. She holds a doctorate in Curriculum and Teaching with an emphasis on Literacy from Teachers College, Columbia University. As a Presidential Management Fellow, Julie worked both on Capitol Hill as an education policy advisor in the U.S. Senate and in the Education Branch in the White House Office of Management and Budget. Prior to her appointment as a PMF, Julie was university faculty teaching courses in writing, linguistics, and Internet ethics. She also served as Technology Director of the Coastal Savannah Writing Project, and has a decade of experience implementing educational technology-centered professional development for teachers across grade levels and subject areas. Julie began her career teaching at the high school level during which time she earned certification from the National Board for Professional Teaching Standards in Adolescence and Young Adulthood English Language Arts. She was a 2013 recipient of the National Council of Teachers of English's Conference on English Leadership's Emerging Leaders Fellowship and was a 2016 International Literacy Association Outstanding Dissertation Award Finalist for her dissertation research on youth and mobile phones.

INDEX

C

D

E

F

G

H

I

J

K

L

M

N

O

P

R

S

new literacies

AND DIGITAL EPISTEMOLOGIES

Colin Lankshear & Michele Knobel
General Editors

New literacies emerge and evolve apace as people from all walks of life engage with new technologies, shifting values and institutional change, and increasingly assume 'postmodern' orientations toward their everyday worlds. Despite many efforts to take account of such changes, educational institutions largely remain out of touch with the range of new ways of making and sharing meanings that increasingly mediate and shape the lives of the young people they teach and the futures they face. This series aims to explore some key dimensions of the changes occurring within social practices of literacy and the educational challenges they present, with a view to informing educational practice in helpful ways. It asks what are new literacies, how do they impact on life in schools, homes, communities, workplaces, sites of leisure, and other key settings of human cultural engagement, and what significance do new literacies have for how people learn and how they understand and construct knowledge. It aims to challenge established and 'official' ways of framing literacy, and to ask what it means for literacies to be powerful, effective, and enabling under current and foreseeable conditions. Collectively, the works in this series will help to reorient literacy debates and literacy education agendas.

For further information about the series and submitting manuscripts, please contact:

Michele Knobel & Colin Lankshear
Montclair State University
Dept. of Education and Human Services
3173 University Hall
Montclair, NJ 07043
michele@coatepec.net

www.ingramcontent.com/pod-product-compliance
Ingram Content Group UK Ltd.
Pitfield, Milton Keynes, MK11 3LW, UK
UKHW020841290926
15470UKWH00004B/1477

9 781433 144073